T0215603

INTRODUCTION TO CONTINUUM MECHANICS

This textbook treats solids and fluids in a balanced manner, using thermodynamic restrictions on the relation between applied forces and material responses. This unified approach can be appreciated by engineers, physicists, and applied mathematicians with some background in engineering mechanics. It has many examples and about 150 exercises for students to practice. The higher mathematics needed for a complete understanding is provided in the early chapters. This subject is essential for engineers involved in experimental or numerical modeling of material behavior.

Sudhakar Nair is the Associate Dean for Academic Affairs of the Graduate College, Professor of Mechanical Engineering and Aerospace Engineering, and Professor of Applied Mathematics at the Illinois Institute of Technology in Chicago. He is a Fellow of the ASME, an Associate Fellow of the AIAA, and a member of the American Academy of Mechanics as well as Tau Beta Pi and Sigma Xi.

Introduction to Continuum Mechanics

Sudhakar Nair
Illinois Institute of Technology

CAMBRIDGE
UNIVERSITY PRESS

CAMBRIDGE UNIVERSITY PRESS
Cambridge, New York, Melbourne, Madrid, Cape Town,
Singapore, São Paulo, Delhi, Tokyo, Mexico City

Cambridge University Press
32 Avenue of the Americas, New York, NY 10013-2473, USA

www.cambridge.org
Information on this title: www.cambridge.org/9780521187893

First published 2009
First paperback edition 2011

A catalog record for this publication is available from the British Library

Library of Congress Cataloging in Publication data

Nair, Sudhakar, 1944–
Introduction to continuum mechanics / Sudhakar Nair.
p. cm.
Includes bibliographical references and index.
ISBN 978-0-521-87562-2 (hardback)
1. Continuum mechanics. I. Title.
QA808.2.N26 2009
531 – dc22 2008043659

ISBN 978-0-521-87562-2 Hardback
ISBN 978-0-521-18789-3 Paperback

Contents

Preface

This text is based on a one-semester course I have been teaching at the Illinois Institute of Technology for about 30 years. Graduate students from mechanical and aerospace engineering, civil engineering, chemical engineering, and applied mathematics have been the main customers. Most of the students in my course have had some exposure to Newtonian fluids and linear elasticity. These two topics are covered here, neglecting the large number of boundary-value problems solved in undergraduate texts. On a number of topics, it becomes necessary to sacrifice depth in favor of breadth, as students specializing in a particular area will be able to delve deeper into that area with the foundation laid out in this course. Space and time constraints prevented the inclusion of classical topics such as hypoelasticity and electromagnetic effects in elastic and fluid materials and a more detailed treatment of nonlinear viscoelastic fluids.

I have included a small selection of exercises at the end of each chapter, and students who attempt some of these exercises will benefit the most from this text. Instructors may add reading assignments from other sources.

Instead of placing all the references at the end of the book, I have given the pertinent books and articles relevant to each chapter at the end of that chapter. There are some duplications in this mode of presentation, but I hope it is more convenient. The Introduction is followed by a list of books on continuum mechanics intended for students looking for deeper insight. Students are cautioned that there are a variety of notations in the literature, and it is recommended that they read the definitions carefully before comparing the equations. In particular, my definition of the deformation gradient happens to be the transverse of that found in many books. The "comma" notation for partial derivatives, which is commonly used in mechanics, usually confuses students when it comes to the order of the indices in a tensor component. I have tried to avoid the "commas" in the beginning while the students are learning to manipulate tensor components.

I am grateful to my colleagues: Michael Gosz, who used this text for a course on continuum mechanics he taught, and Hassan Nagib and Candace Wark for trying out the chapter on Cartesian tensors on their fluid dynamics students. I am also indebted to Professors Sia Nemat-Nasser and Gil Hegemier for instilling in me an appreciation for continuum mechanics. A number of students in my classes took it upon themselves to prepare detailed errata for the notes I had distributed. I appreciate their input and have tried to incorporate all of their corrections (while, probably, adding some new errors).

The editorial help provided by Peter Gordon, Vicki Danahy, Barbara Walthall, and Cambridge University Press is highly appreciated.

Throughout the preparation of the manuscript my wife, Celeste, has provided constant encouragement, and I am thankful to her.

Introduction to Continuum Mechanics

1 Introduction

Mechanics is the study of the behavior of matter under the action of internal and external forces. In this introductory treatment of continuum mechanics, we accept the concepts of time, space, matter, energy, and force as the Newtonian ideals. Here our objective is the formulation of engineering problems consistent with the fundamental principles of mechanics. To paraphrase Professor Y. C. Fung—there are generally two ways of approaching mechanics: One is the ad hoc method, in which specific problems are considered and specific solution methods are devised that incorporate simplifying assumptions, and the other is the general approach, in which the general features of a theory are explored and specific applications are considered at a later stage. Engineering students are familiar with the former approach from their experience with "Strength of Materials" in the undergraduate curriculum. The latter approach enables them to understand an entire field in a systematic way in a short time. It has been traditional, at least in the United States, to have a course in continuum mechanics at the senior or graduate level to unify the ad hoc concepts students have learned in the undergraduate courses. Having had the knowledge of thermodynamics, fluid dynamics, and strength of materials, at this stage, we look at the entire field in a unified way.

1.1 Concept of a Continuum

Although mechanics is a branch of physics in which, according to current developments, space and time may be discrete, in engineering the length and time scales are orders (and orders) of magnitude larger than those in quantum physics and we use space coordinates and time as continuous. The concept of a continuum refers to the treatment of matter as continuous. The justification for this, again, rests on the length scales involved. For example, consider a large volume V of air under constant pressure and temperature. Within this volume, visualize a small volume ΔV centered at a fixed point in space. Let us denote by ΔM the mass of material inside ΔV. The ratio $\Delta M/\Delta V$ is the average density ρ. However, if we shrink ΔV, we can imagine a state in which molecules of oxygen and nitrogen pass through it, and the concept of density loses its meaning. If the dimension of ΔV is kept large

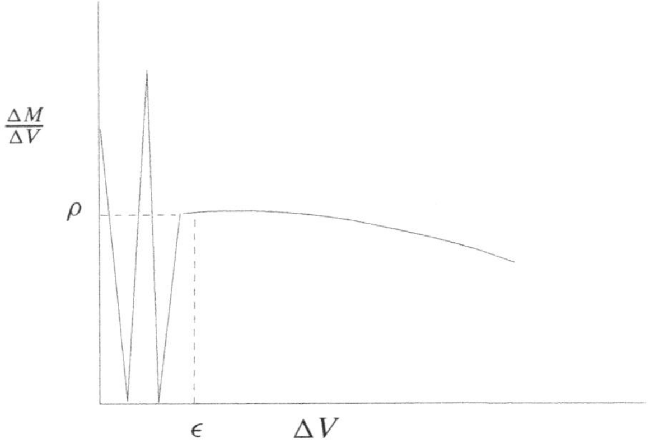

Figure 1.1. Density variation with volume (based on Fung, 1993).

compared with the mean free path of the molecules, we could use the concept of density. Mathematically, we say, "let $\Delta V \to 0$," but, physically, it is still kept above some value $\epsilon \gg 0$. We may also think of the situation as the discrete particles of matter are approximated as a continuous "smeared" state. Figure 1.1 shows a sketch of the limiting process.

To distinguish moving matter from its fixed-background space we use the term **material particle**, which is not to be confused with molecules or atoms.

1.2 Sequence of Topics

Prior to the consideration of mechanics topics such as stress and strain, the mathematical apparatus needed for our work is briefly reviewed. The fixed space in which the continuum moves is a three-dimensional (3D) Euclidean space. Cartesian tensors are essential for describing the deformation, motion, and the forces in mechanics within a Cartesian coordinate system. This topic is considered in Chapter 2. Although general tensors are not required for our studies, an understanding of this topic is worthwhile to appreciate its connection to geometry and mechanics. A number of advanced works in continuum mechanics use the general tensor formulation that is invariant under coordinate transformations involving curvilinear coordinates. Chapter 3 introduces some of the basic properties of general tensors.

Integral theorems of calculus, namely, the theorems of Gauss, Green, and Stokes, are extremely useful for our study of mechanics. These theorems are known to students from their studies of calculus. In Chapter 4 we visit these theorems by using index notation.

The next two chapters, Chapter 5 and Chapter 6, deal with the descriptions of the geometry of deforming bodies. Various strain measures and strain rate quantities are introduced in these chapters.

A separate chapter, Chapter 7, is devoted to a discussion of the fundamental axioms of mechanics. For a proper introduction of the stress tensor in Chapter 8, the fundamental axiom dealing with the balance of momentum is essential.

Thermodynamics plays an important role not only in restricting the form of stress–strain relations and stress–strain rate relations but also in explaining thermomechanical coupling. Chapter 9 refreshes the readers' knowledge in thermodynamics.

General forms of constitutive relations and their admissible forms are discussed in Chapter 10.

Chapter 11 considers elastic materials. It starts with nonlinear elastic materials, presents some of the classic inverse solutions, and then proceeds to linear elastic materials. A section on rubber elasticity is included because of the importance of this topic in engineering applications.

Chapter 12 deals with fluid dynamics. Again, classic inverse solutions of nonlinear fluids are presented first. Newtonian fluids and Navier–Stokes equations are briefly discussed.

Most students are unfamiliar with viscoelasticity and plasticity. These two topics are dealt with in Chapters 13 and 14. The treatment is of an elementary nature as this is assumed to be the first exposure of these two topics.

A digression into the available numerical solution techniques applicable to nonlinear problems is not made. The finite-element method has become the method of choice to deal with the solutions of solid mechanics problems. Finite-difference methods are often used in fluid dynamics problems. Other methods, such as molecular dynamics and Monte Carlo methods (see Frenkel and Smit, 2002), are actually outside the domain of continuum mechanics. However, these two methods play crucial roles in illuminating the foundations of irreversible thermodynamics (see Yourgrau, van der Merwe, and Raw, 1966) and continuum mechanics. Students are encouraged to supplement the topics covered here with courses in statistical mechanics (see Chandler, 1987) and related numerical methods.

SUGGESTED READING

Batchelor, G. K. (1967). *An Introduction to Fluid Dynamics*, Cambridge University Press.

Batra, R. (2005). *Elements of Continuum Mechanics*, AIAA Publishers.

Chandler, D. (1987). *Introduction to Modern Statistical Mechanics*, Oxford University Press.

Eringen, A. C. (1962). *Nonlinear Theory of Continuous Media*, McGraw-Hill.

Frenkel, D. and Smit, B. (2002). *Understanding Molecular Simulation*, Academic.

Fung, Y. C. (1965). *Foundations of Solid Mechanics*, Prentice-Hall.

Fung, Y. C. (1993). *First Course in Continuum Mechanics*, 3rd ed., Prentice-Hall.

Jaunzemis, W. (1967). *Continuum Mechanics*, Macmillan.

Malvern, L. E. (1969). *Introduction to the Mechanics of a Continuous Medium*, Prentice-Hall.

Reddy, J. N. (2008). *An Introduction to Continuum Mechanics.* Cambridge University Press.

Truesdell, C. (1960). *Principles of Continuum Mechanics*, Field Research Laboratory, Socony Mobil Oil Co., Dallas, TX.

Truesdell, C. and Noll, W. (1965). The nonlinear field theories of mechanics, in *Encyclopedia of Physics* (S. Flügge, ed.), Springer-Verlag, Vol. 3/3.

Truesdell, C. and Toupin, R. A. (1960). The classical field theories, in *Encyclopedia of Physics* (S. Flügge, ed.), Springer-Verlag, Vol. 3/1.

Yourgrau, W., van der Merwe A., and Raw, G. (1966). *Treatise on Irreversible and Statistical Thermodynamics*, Macmillan.

2 Cartesian Tensors

When the coordinates used to describe the geometry and deformation of a continuum and the forces involved are Cartesian, that is, three mutually orthogonal, right-handed coordinates with the Euclidean formula for distances, the quantities entering the equations of motion are conveniently described by use of the Cartesian tensors. Before we familiarize ourselves with these, let us examine a few related topics. These topics are included here primarily to establish our notation and to refresh the concepts the students might have seen in other contexts.

2.1 Index Notation and Summation Convention

Index notation uses coordinates x_1, x_2, and x_3 to denote the classical x, y, and z coordinates, respectively. The components of a vector $\boldsymbol{v}$ would be v_1, v_2, and v_3 (in three dimensions), instead of the conventional u, v, and w. As far as matrix elements are concerned, index notation, such as A_{23} to identify the element in the second row and third column, has been in use for some time. The advantage of index notation, in conjunction with the summation convention, is that we can shorten long mathematical expressions.

Consider a system of M equations, in N unknowns:

$$\begin{aligned} A_{11}x_1 + A_{12}x_2 + \cdots + A_{1N}x_N &= c_1, \\ A_{21}x_1 + A_{22}x_2 + \cdots + A_{2N}x_N &= c_2, \\ \cdots\cdots\cdots\cdots\cdots\cdots\cdots\cdots &= , \ldots, \\ A_{M1}x_1 + A_{M2}x_2 + \cdots + A_{MN}x_N &= c_M. \end{aligned} \tag{2.1}$$

This system of equations can also be written as

$$\sum_{j=1}^{N} A_{ij}x_j = c_i \quad (i = 1, 2, \ldots, M; j = 1, 2, \ldots, N).$$

In accordance with the **Einstein summation convention** we can further simplify the notation by writing

$$A_{ij}x_j = c_i \; (i = 1, 2, \ldots, M; j = 1, 2, \ldots, N), \tag{2.2}$$

where summation on the repeated index j is implied. Here, i is called a free index and j is called a dummy index. In dealing with three-dimensional (3D) Euclidean space, we have indices ranging from 1 to 3 (i.e., $M = N = 3$). Whenever an index is repeated once (and only once), the terms have to be summed with respect to that index. For this convention to be effective, extreme care must be taken to avoid the occurrence of any index more than twice. As examples, we have

$$A_{ii} = A_{11} + A_{22} + A_{33},$$

$$A_{ij} B_{ij} = B_{ij} A_{ij} = B_{ji} A_{ji}.$$

The symmetry of an array can be expressed as

$$A_{ij} = A_{ji}. \tag{2.3}$$

If an array is *skew symmetric* (also called antisymmetric),

$$B_{ij} = -B_{ji}. \tag{2.4}$$

An arbitrary array C can be expressed as the sum of a symmetric and a *skew-symmetric* array:

$$C_{ij} = A_{ij} + B_{ij}, \tag{2.5}$$

where

$$A_{ij} = C_{(ij)} = \frac{1}{2}(C_{ij} + C_{ji}), \quad B_{ij} = C_{[ij]} = \frac{1}{2}(C_{ij} - C_{ji}). \tag{2.6}$$

The subscripts inside the parentheses and square brackets help us to avoid introducing new variables A and B.

There are rare occasions when we would like to suppress the summation convention. Suppose we want to refer to A_{11}, A_{22}, or A_{33}; if we use A_{ii} we get the sum of the three terms. We may underline the repeated index to suppress the summation:

$$A_{\underline{ii}} = A_{11},\ A_{22},\ \text{or}\ A_{33}. \tag{2.7}$$

When we need to substitute one formula into another, we have to make sure that the dummy indices are distinct. For example,

$$a_i = C_{ij} b_j, \quad b_j = c_i D_{ji}. \tag{2.8}$$

A direct substitution for b_j in the first equation shows

$$a_i = C_{ij} c_i D_{ji}, \tag{2.9}$$

where i appears three times on the right-hand side, violating the summation rule. To avoid this, first we write

$$b_j = c_k D_{jk}, \tag{2.10}$$

and then substitute in the first equation to get

$$a_i = C_{ij} c_k D_{jk} = C_{ij} D_{jk} c_k, \tag{2.11}$$

where j and k are summation indices (or dummy indices) and the free index i appears on both sides. The dummy indices are similar to dummy variables used in integration of functions.

2.2 Kronecker Delta and Permutation Symbol

These two notations are extremely useful in connection with **Cartesian coordinates**. The Kronecker delta is defined as

$$\delta_{ij} = \begin{cases} 1, & \text{if} \quad i = j \\ 0, & \text{if} \quad i \neq j \end{cases}. \tag{2.12}$$

This definition assumes that i and j are explicit integers, such as $i = 1$ and $j = 3$, and it does not imply $\delta_{ii} = 1$. Elements of the Kronecker delta correspond to the elements of the identity matrix

$$\boldsymbol{I} = [\delta_{ij}] = \begin{bmatrix} 1 & 0 & 0 \\ 0 & 1 & 0 \\ 0 & 0 & 1 \end{bmatrix}. \tag{2.13}$$

With this,

$$\delta_{ii} = 3, \quad \delta_{ij}\delta_{jk} = \delta_{ik}, \quad \delta_{ij} A_{jk} = A_{ik}. \tag{2.14}$$

The permutation symbol or alternator is defined as

$$e_{ijk} = \begin{cases} 1, & \text{if} \quad i, j, k \text{ are even permutations of } 1, 2, 3 \\ -1, & \text{if} \quad i, j, k \text{ are odd permutations of } 1, 2, 3 \\ 0, & \text{otherwise.} \end{cases} \tag{2.15}$$

From the preceding definition we have

$$e_{ijk} = e_{jki} = e_{kij} = -e_{ikj} = -e_{jik} = -e_{kji}, \tag{2.16}$$

$$e_{ijj} = e_{jij} = e_{jji} = 0. \tag{2.17}$$

Explicitly we have

$$e_{123} = e_{231} = e_{312} = 1, \quad e_{213} = e_{321} = e_{132} = -1. \tag{2.18}$$

Figure 2.1 shows the order of the indices for even and odd permutations.

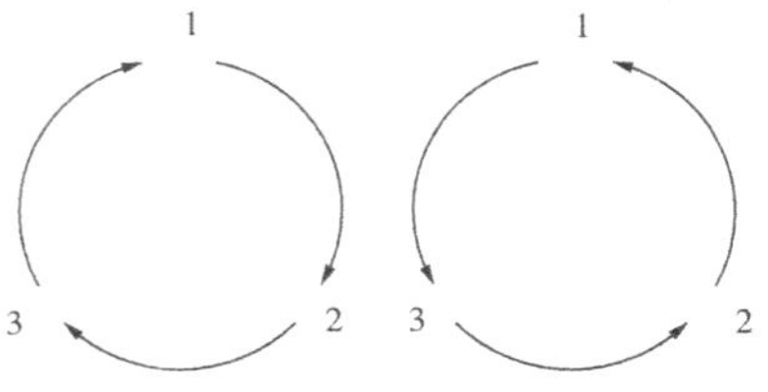

Figure 2.1. Even and odd permutations of the integers 1, 2, and 3.

Using the permutation symbol, we may express the determinant of a 3×3 – matrix $\boldsymbol{A}$ as

$$\begin{aligned}\det A = |\boldsymbol{A}| &= A_{11}A_{22}A_{33} + A_{12}A_{23}A_{31} + A_{13}A_{21}A_{32} \\ &\quad - A_{11}A_{23}A_{32} - A_{12}A_{21}A_{33} - A_{13}A_{22}A_{31} \\ &= e_{ijk}A_{1i}A_{2j}A_{3k} \\ &= e_{ijk}A_{i1}A_{j2}A_{k3} \\ &= \tfrac{1}{6}e_{ijk}e_{\ell mn}A_{i\ell}A_{jm}A_{kn}.\end{aligned} \tag{2.19}$$

We can see the second and third equations as the (first) row expansion and the (first) column expansion of the determinant and the last equation as the sum of all row expansions and all column expansions (which add up to 6), divided by 6.

We can also express the determinant in terms of the cofactors of the matrix. For a 3×3 matrix the cofactor of an element A_{ij} is the 2×2 determinant we obtain by eliminating the ith row and jth column and multiplying it by $(-1)^{i+j}$. Let us denote this cofactor by A^*_{ij}. Then

$$|\boldsymbol{A}| = A_{ij}A^*_{ij} \quad \text{(no sum on } i\text{)}. \tag{2.20}$$

Observing that A^*_{ij} does not contain A_{ij} itself (recall, we eliminated a row and a column), we find

$$A^*_{ij} = \frac{\partial |A|}{\partial A_{ij}}. \tag{2.21}$$

When the inverse of the matrix exists, we have

$$A^{-1}_{ij} = \frac{A^*_{ji}}{|A|} = \frac{1}{|A|}\frac{\partial |A|}{\partial A_{ji}}. \tag{2.22}$$

A relation between the permutation symbols and the Kronecker deltas, known as the "e–δ identity," is useful in algebraic simplifications:

$$e_{ijk}e_{mnk} = \delta_{im}\delta_{jn} - \delta_{in}\delta_{jm}. \tag{2.23}$$

From this, when $n = j$, we get

$$e_{ijk}e_{mjk} = 2\delta_{im}, \tag{2.24}$$

and, further, when $m = i$,

$$e_{ijk}e_{ijk} = 6. \tag{2.25}$$

2.2.1 Example: Skew Symmetry

If A_{ij} is a skew-symmetric matrix, solve the system of equations

$$e_{ijk}A_{jk} = B_i. \tag{2.26}$$

We use the "e–δ" identity to get

$$
\begin{aligned}
e_{imn}e_{ijk}A_{jk} &= e_{imn}B_i, \\
(\delta_{mj}\delta_{nk} - \delta_{mk}\delta_{nj})A_{jk} &= e_{imn}B_i, \\
A_{mn} - A_{nm} &= e_{imn}B_i, \\
A_{mn} &= \frac{1}{2}e_{imn}B_i. \qquad (2.27)
\end{aligned}
$$

where we use the skew-symmetry property $A_{nm} = -A_{mn}$.

2.2.2 Example: Products

If A_{ij} is symmetric and B_{ij} is skew-symmetric, show that $A_{ij}B_{ij} = 0$.

Let

$$S = A_{ij}B_{ij}. \qquad (2.28)$$

If we interchange the dummy indices $i \to j$ and $j \to i$, we have

$$S = A_{ji}B_{ji}. \qquad (2.29)$$

Using the symmetry of A and the skew symmetry of B, we can write this as

$$S = A_{ij}(-B_{ij}) = -A_{ij}B_{ij} = -S, \quad 2S = 0, \quad S = 0. \qquad (2.30)$$

2.3 Coordinate System

As shown in Fig. 2.2, we use a proper (right-handed) Cartesian coordinate system with x_1, x_2, and x_3 denoting the three axes. A directed line segment from the origin to any point in this 3D space is called a position vector $\boldsymbol{r}$, with components x_1, x_2,

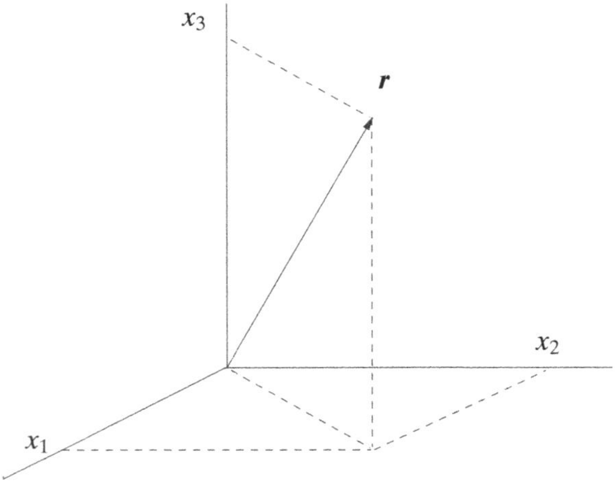

Figure 2.2. Cartesian coordinate system.

and x_3 along the three axes. We also use the notation $\boldsymbol{x}$ instead of $\boldsymbol{r}$ at times. In matrix notation, using a column vector, we have

$$\boldsymbol{r} = \boldsymbol{x} = \begin{Bmatrix} x_1 \\ x_2 \\ x_3 \end{Bmatrix}. \tag{2.31}$$

Defining unit vectors (or base vectors) $\boldsymbol{e}_1$, $\boldsymbol{e}_2$, and $\boldsymbol{e}_3$ as

$$\boldsymbol{e}_1 = \begin{Bmatrix} 1 \\ 0 \\ 0 \end{Bmatrix}, \quad \boldsymbol{e}_2 = \begin{Bmatrix} 0 \\ 1 \\ 0 \end{Bmatrix}, \quad \boldsymbol{e}_3 = \begin{Bmatrix} 0 \\ 0 \\ 1 \end{Bmatrix}, \tag{2.32}$$

we can write

$$\boldsymbol{r} = x_1\boldsymbol{e}_1 + x_2\boldsymbol{e}_2 + x_3\boldsymbol{e}_3. \tag{2.33}$$

The **dot** and **cross products** of the unit vectors can be expressed with the Kronecker delta and the permutation symbol in the form

$$\boldsymbol{e}_i \cdot \boldsymbol{e}_j = \delta_{ij}, \quad \boldsymbol{e}_i \times \boldsymbol{e}_j = e_{ijk}\boldsymbol{e}_k. \tag{2.34}$$

2.4 Coordinate Transformations

Two types of coordinate transformations are encountered frequently in our studies: coordinate translation and coordinate rotation. If we denote the new coordinates of a point $\boldsymbol{P}$ by x_i', in the case of translation, the two systems are related in the form

$$x_i' = x_i + h_i, \tag{2.35}$$

where h_i are constants. This is shown in Fig. 2.3.

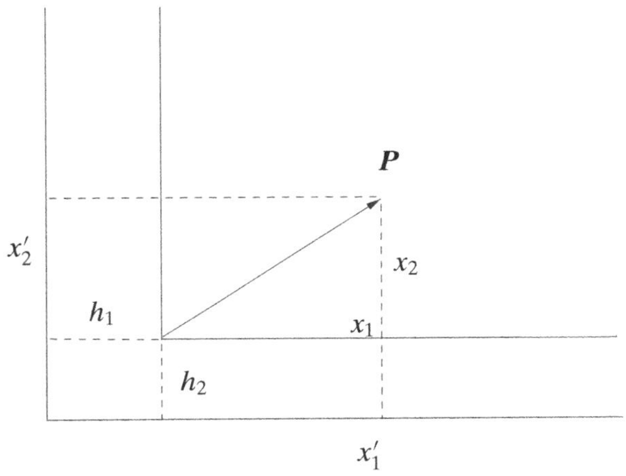

Figure 2.3. Coordinate translation.

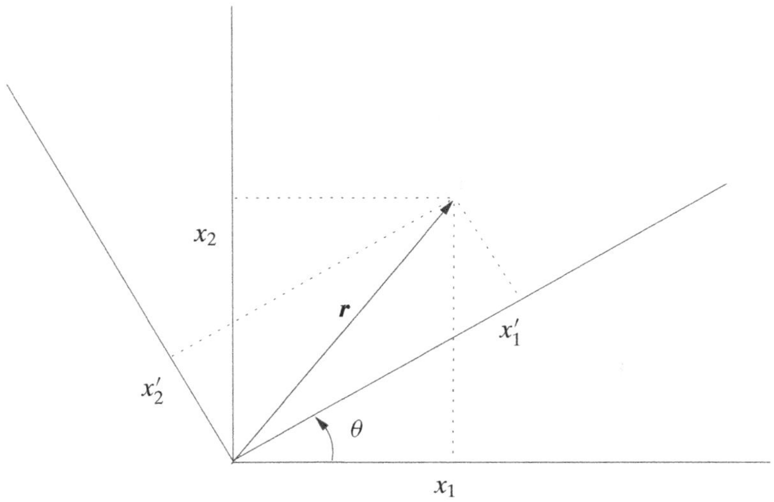

Figure 2.4. Coordinate rotation.

Let us consider first a rotation of coordinates in two dimensions. The origin remains fixed and we obtain the x'_1 axis by rotating the x_1 axis counterclockwise by an angle, θ. As shown in Fig. 2.4, the two systems are related in the form

$$\begin{aligned} x'_1 &= x_1 \cos\theta + x_2 \sin\theta, \\ x'_2 &= -x_1 \sin\theta + x_2 \cos\theta. \end{aligned} \tag{2.36}$$

We obtain the inverse of this transformation by replacing θ with $(-\theta)$, as

$$\begin{aligned} x_1 &= x'_1 \cos\theta - x'_2 \sin\theta, \\ x_2 &= x'_1 \sin\theta + x'_2 \cos\theta. \end{aligned}$$

Equations (2.36) can also be written in matrix form:

$$\begin{Bmatrix} x'_1 \\ x'_2 \end{Bmatrix} = \begin{bmatrix} Q_{11} & Q_{12} \\ Q_{21} & Q_{22} \end{bmatrix} \begin{Bmatrix} x_1 \\ x_2 \end{Bmatrix}, \tag{2.37}$$

where

$$Q_{11} = Q_{22} = \cos\theta, \quad Q_{12} = -Q_{21} = \sin\theta. \tag{2.38}$$

Using column vectors $\boldsymbol{x}'$ and $\boldsymbol{x}$ and the square matrix $\boldsymbol{Q}$, we have

$$\boldsymbol{x}' = \boldsymbol{Q}\boldsymbol{x}. \tag{2.39}$$

A note of caution is in order at this point. We have absolute vectors in space, such as $\boldsymbol{r}$, and then we have column representations of the components in a chosen coordinate system, such as $\boldsymbol{x}$ and $\boldsymbol{x}'$. When there are no coordinate rotations involved, we do not have to distinguish these two representations. When there are coordinate rotations, we use $\boldsymbol{r}$ for the absolute vector.

Another way to relate $\boldsymbol{x}'$ to $\boldsymbol{x}$ is by use of two systems of unit vectors, $\boldsymbol{e}_i'$ and $\boldsymbol{e}_i$. Let us do this for the 3D case. The position vector $\boldsymbol{r}$ has two representations:

$$\boldsymbol{r} = x_1'\boldsymbol{e}_1' + x_2'\boldsymbol{e}_2' + x_3'\boldsymbol{e}_3' = x_1\boldsymbol{e}_1 + x_2\boldsymbol{e}_2 + x_3\boldsymbol{e}_3. \tag{2.40}$$

Because $\boldsymbol{e}_1'$, $\boldsymbol{e}_2'$, and $\boldsymbol{e}_3'$ are mutually orthogonal, taking the dot product of the preceding relation with $\boldsymbol{e}_1'$ gives

$$x_1' = (\boldsymbol{e}_1' \cdot \boldsymbol{e}_1)x_1 + (\boldsymbol{e}_1' \cdot \boldsymbol{e}_2)x_2 + (\boldsymbol{e}_1' \cdot \boldsymbol{e}_3)x_3. \tag{2.41}$$

The elements of the 3×3 $\boldsymbol{Q}$ matrix associated with the rotation are the dot products of the unit vectors:

$$Q_{11} = \boldsymbol{e}_1' \cdot \boldsymbol{e}_1, \quad Q_{12} = \boldsymbol{e}_1' \cdot \boldsymbol{e}_2, \quad Q_{13} = \boldsymbol{e}_1' \cdot \boldsymbol{e}_3. \tag{2.42}$$

Next, let us do this using the index notation and the summation convention:

$$\boldsymbol{r} = x_k'\boldsymbol{e}_k' = x_j\boldsymbol{e}_j. \tag{2.43}$$

Dotting with $\boldsymbol{e}_i'$, we obtain

$$x_k'\boldsymbol{e}_i' \cdot \boldsymbol{e}_k' = x_j\boldsymbol{e}_i' \cdot \boldsymbol{e}_j, \quad x_k'\delta_{ik} = Q_{ij}x_j,$$

or

$$x_i' = Q_{ij}x_j, \quad \boldsymbol{x}' = \boldsymbol{Q}\boldsymbol{x}, \tag{2.44}$$

where

$$Q_{ij} = \boldsymbol{e}_i' \cdot \boldsymbol{e}_j. \tag{2.45}$$

We may start with Eq. (2.43) and dot it with $\boldsymbol{e}_i$. This would lead to

$$x_i = \boldsymbol{e}_i \cdot \boldsymbol{e}_j' x_j'. \tag{2.46}$$

This is the inverse of Eqs. (2.44):

$$x_i = Q_{ji}x_j' = Q_{ij}^T x_j', \quad \boldsymbol{x} = \boldsymbol{Q}^T\boldsymbol{x}'. \tag{2.47}$$

Using Eqs. (2.47) in Eqs. (2.44), we find

$$\boldsymbol{x}' = \boldsymbol{Q}\boldsymbol{Q}^T\boldsymbol{x}', \tag{2.48}$$

where the superscript T denotes transpose. This shows

$$\boldsymbol{Q}\boldsymbol{Q}^T = \boldsymbol{I} \quad \text{or} \quad \boldsymbol{Q}^{-1} = \boldsymbol{Q}^T. \tag{2.49}$$

Matrices that satisfy this relation are called orthogonal matrices. Another way to see that $\boldsymbol{Q}$ is orthogonal is as follows:

The length of the vector $\boldsymbol{r}$ can be found from

$$r^2 = x_i'x_i' = x_jx_j. \tag{2.50}$$

Substituting for x_i', we obtain

$$Q_{im}Q_{in}x_m x_n = x_i x_i \quad \text{or} \quad x_m Q_{mi}^T Q_{in} x_n = x_i x_i. \tag{2.51}$$

We can rewrite this in matrix notation as

$$\boldsymbol{x}^T \boldsymbol{Q}^T \boldsymbol{Q}\boldsymbol{x} = \boldsymbol{x}^T \boldsymbol{x} \quad \text{or} \quad \boldsymbol{Q}^T \boldsymbol{Q} = \boldsymbol{I}, \tag{2.52}$$

$$Q_{ik}^T Q_{kj} = \delta_{ij} \quad \text{or} \quad Q_{ki} Q_{kj} = \delta_{ij}, \tag{2.53}$$

where we have used the fact that $\boldsymbol{x}$ is arbitrary. Thus

$$\boldsymbol{Q}^T = \boldsymbol{Q}^{-1}. \tag{2.54}$$

As $Q_{ij} \neq Q_{ji}$, we associate the first index i with the $\boldsymbol{e}'$ system and j with the $\boldsymbol{e}$ system.

Also, the unit vectors $\boldsymbol{e}_i'$ are related to $\boldsymbol{e}_j$ in the form

$$\boldsymbol{e}_i' = Q_{ij}\boldsymbol{e}_j, \tag{2.55}$$

which we can verify by taking the dot product with $\boldsymbol{e}_k$.

2.5 Vectors

The mathematical definition of vectors as directed line segments with their starting point fixed at the origin enables us to express any vector $\boldsymbol{v}$ in a chosen coordinate system as

$$\boldsymbol{v} = v_i \boldsymbol{e}_i, \tag{2.56}$$

where v_i are the components of the vector in that system.

In another system with unit vectors $\boldsymbol{e}'$, we will have

$$\boldsymbol{v} = v_i' \boldsymbol{e}_i'. \tag{2.57}$$

From our examination of the coordinate transformation, it is clear that the two sets of components are related in the form

$$v_i' = Q_{ij} v_j. \tag{2.58}$$

The dot product (inner product or scalar product) of two vectors $\boldsymbol{u}$ and $\boldsymbol{v}$ is defined as

$$\boldsymbol{u} \cdot \boldsymbol{v} = u_i \boldsymbol{e}_i \cdot v_j \boldsymbol{e}_j = u_i v_j \boldsymbol{e}_i \cdot \boldsymbol{e}_j = u_i v_j \delta_{ij} = u_i v_i. \tag{2.59}$$

The cross product (vector product) of these vectors is defined as

$$\boldsymbol{u} \times \boldsymbol{v} = u_i \boldsymbol{e}_i \times v_j \boldsymbol{e}_j = u_i v_j \boldsymbol{e}_i \times \boldsymbol{e}_j = u_i v_j e_{ijk} \boldsymbol{e}_k. \tag{2.60}$$

In physics and engineering there are many 3×1 arrays that transform from one coordinate system to another according to the transformation rule of Eq. (2.58). We call them vectors. Examples include position vectors, velocities, forces (relative to their point of application), and gradients of scalar functions. That brings us to scalars. These are quantities with no associated directions. Examples include temperature, hydrostatic pressure, and density of matter.

A scalar function ψ can be written as

$$\psi = \psi(x_1, x_2, x_3). \tag{2.61}$$

In a different coordinate system, using

$$x_i = Q_{ji}x'_j, \tag{2.62}$$

we obtain

$$\psi(x_1, x_2, x_3) = \psi(Q_{j1}x'_j, Q_{j2}x'_j, Q_{j3}x'_j) = \psi'(x'_1, x'_2, x'_3), \tag{2.63}$$

which shows that the functional forms of ψ and ψ' are different. The numerical value of the function at a given point obtained from $\psi(x'_1, x'_2, x'_3)$ or $\psi(x_1, x_2, x_3)$ will be the same.

When we examine the transformation of the derivatives of a scalar function, we use the chain rule of differentiation to get

$$\frac{\partial \psi'}{\partial x'_i} = \frac{\partial \psi}{\partial x_j}\frac{\partial x_j}{\partial x'_i}. \tag{2.64}$$

From Eq. (2.62),

$$\frac{\partial x_j}{\partial x'_i} = Q_{ij}. \tag{2.65}$$

Thus,

$$\frac{\partial \psi'}{\partial x'_i} = \frac{\partial \psi}{\partial x_j} Q_{ij}. \tag{2.66}$$

So the array formed by the three derivatives of ψ obeys the transformation rule for a vector. We define the gradient vector as

$$\nabla\psi = \begin{Bmatrix} \dfrac{\partial \psi}{\partial x_1} \\ \dfrac{\partial \psi}{\partial x_2} \\ \dfrac{\partial \psi}{\partial x_3} \end{Bmatrix} = \boldsymbol{e}_1\frac{\partial \psi}{\partial x_1} + \boldsymbol{e}_2\frac{\partial \psi}{\partial x_2} + \boldsymbol{e}_3\frac{\partial \psi}{\partial x_3}. \tag{2.67}$$

We can also introduce a vector operator,

$$\nabla = \boldsymbol{e}_i \partial_i, \tag{2.68}$$

where

$$\partial_i = \frac{\partial}{\partial x_i}. \tag{2.69}$$

In the x' system, we could use

$$\partial'_i = \frac{\partial}{\partial x'_i}. \tag{2.70}$$

To simplify our labor, we can express functions of a set of independent variables that belongs to an unambiguous family by using the condensed form

$$\psi(\boldsymbol{x}) = \psi(x_1, x_2, x_3). \tag{2.71}$$

2.6 Tensors

In 3D space, the nine quantities T_{ij} are called the components of a tensor $\boldsymbol{T}$ of rank 2 if they obey the transformation rule,

$$T'_{ij} = Q_{im} Q_{jn} T_{mn}, \tag{2.72}$$

when the two coordinate systems are related by the rotation

$$x'_i = Q_{ij} x_j.$$

The number of indices shows the rank of a tensor. We call $\boldsymbol{T}$ a second-rank tensor. In three dimensions a second-rank tensor has 3^2 components, and in two dimensions it has 2^2 components.

We can introduce a tensor of rank r in three dimensions with 3^r components,

$$T_{i_1 i_2 \ldots i_r}. \tag{2.73}$$

The transformation rule for these components will have r number of Q's in the form

$$T'_{i_1 i_2 \ldots i_r} = Q_{i_1 j_1} Q_{i_2 j_2} \cdots Q_{i_r j_r} T_{j_1 j_2 \ldots j_r}. \tag{2.74}$$

From a heuristic definition of vectors as directed line segments in space, it is clear that these are independent of the coordinate system. However, to define the components of a vector we need a frame of reference. If we know the representation of a vector in one coordinate system, we can use the transformation rule to find them in any other system. Similarly, the tensor $\boldsymbol{T}$ is an absolute quantity, and its component representation depends on the coordinate system. Again, when we know its components in one system, we use the transformation rule to obtain them in another system of coordinates.

In continuum mechanics, most of the time, we deal with 3D tensors of rank 2. From the relation between rank r and the number of Q factors in the transformation, it is clear that vectors are tensors of rank 1 and scalars are those of rank 0. In our studies we refer to these as vectors or scalars, and we reserve the term "tensor" for tensors of rank 2 and higher.

When we encounter an array of 3^r elements, without examining how they transform from one coordinate system to another, we cannot conclude we have a tensor of rank r. We often encounter various arrays in spreadsheets (for example, homework grades); these are, obviously, not tensors!

2.6.1 Examples of Tensors

In this subsection we examine three examples of tensors.

We may construct second-rank tensors from vectors by using the idea of tensor products. We have seen the dot product and the cross product of two vectors. The tensor product of two vectors $\boldsymbol{u}$ and $\boldsymbol{v}$ is written as

$$\boldsymbol{u} \otimes \boldsymbol{v} = u_i \boldsymbol{e}_i \otimes v_j \boldsymbol{e}_j = u_i v_j \boldsymbol{e}_i \otimes \boldsymbol{e}_j. \tag{2.75}$$

This operation produces a 3×3 array, with the coefficient of, say, $\boldsymbol{e}_2 \otimes \boldsymbol{e}_3$ entering the second row and third column. Let

$$\boldsymbol{w} = \boldsymbol{u} \otimes \boldsymbol{v}, \qquad w_{ij} = u_i v_j. \tag{2.76}$$

In a new coordinate system,

$$w'_{ij} = u'_i v'_j = Q_{im} u_m Q_{jn} v_n = Q_{im} Q_{jn} w_{mn}. \tag{2.77}$$

Thus the components of $\boldsymbol{w}$ obey the transformation law for second-rank tensors. In our studies, we use the dyadic notation that implicitly assumes the tensor product symbol, $\otimes$, between two vectors if a dot or cross product symbol is absent. That is,

$$\boldsymbol{uv} = u_i v_j \boldsymbol{e}_i \boldsymbol{e}_j = u_i v_j \boldsymbol{e}_i \otimes \boldsymbol{e}_j = \boldsymbol{u} \otimes \boldsymbol{v}. \tag{2.78}$$

The transpose of the tensor $\boldsymbol{w}$ can be written as

$$\boldsymbol{w}^T = w_{ij}^T \boldsymbol{e}_i \boldsymbol{e}_j = w_{ji} \boldsymbol{e}_i \boldsymbol{e}_j = w_{ij} \boldsymbol{e}_j \boldsymbol{e}_i. \tag{2.79}$$

In matrix notation, if we interpret $\boldsymbol{u}$ and $\boldsymbol{v}$ as column vectors, the tensor product gives the matrix

$$\boldsymbol{w} = \boldsymbol{uv}^T. \tag{2.80}$$

For tensors of a rank higher than 2, we do not have this type of correspondence with matrices. Most of the operations with second-rank tensors we encounter can be carried out in matrix notation.

A second example of constructing tensors is given by use of the gradient operator $\boldsymbol{\nabla}$. We have

$$\boldsymbol{\nabla\nabla}\psi = \boldsymbol{e}_i \partial_i \boldsymbol{e}_j \partial_j \psi = \boldsymbol{e}_i \boldsymbol{e}_j \frac{\partial^2 \psi}{\partial x_i \partial x_j}. \tag{2.81}$$

For a vector-valued function $\boldsymbol{\phi}$, we have

$$\begin{aligned} \boldsymbol{\phi} &= \boldsymbol{e}_i \phi_i, & \\ \boldsymbol{\nabla} \cdot \boldsymbol{\phi} &= \partial_i \phi_i & \text{(divergence)}, \\ \boldsymbol{\nabla} \times \boldsymbol{\phi} &= \partial_i \phi_j \, e_{ijk} \boldsymbol{e}_k & \text{(curl)}, \\ \boldsymbol{\nabla\phi} &= \partial_i \phi_j \, \boldsymbol{e}_i \boldsymbol{e}_j & \text{(gradient)}. \end{aligned} \tag{2.82}$$

In fluid dynamics the velocity vector $\boldsymbol{v}$, its divergence $\boldsymbol{\nabla} \cdot \boldsymbol{v}$, its curl $\boldsymbol{\nabla} \times \boldsymbol{v}$, and its gradient $\boldsymbol{\nabla v}$ all play significant roles.

As a third example, we consider the stress tensor $\boldsymbol{\sigma}$. In French, the word *tension* means stress and the word *tensor* originated from *tension* to denote a set of quantities that transforms as stress components under a coordinate transformation. For simplicity, let us consider the classical force balance of a triangle OAB, as shown in

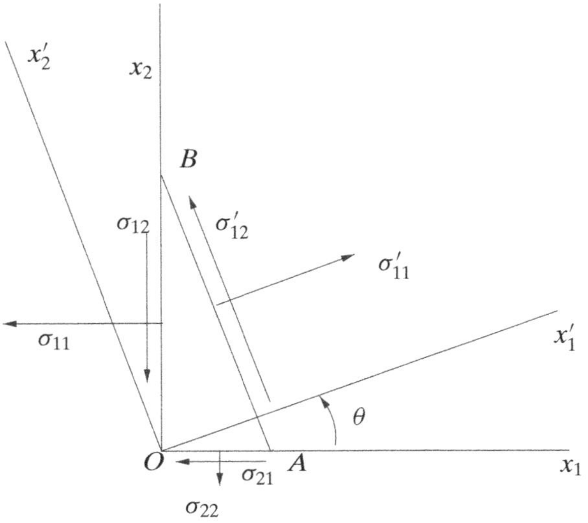

Figure 2.5. Components of stress in two dimensions.

Fig. 2.5. With $OA = AB \sin\theta$ and $OB = AB \cos\theta$, the stress components σ_{11}, σ_{12}, σ_{21}, and σ_{22} in the x_1, x_2 coordinate system are related to the stresses $\sigma_{11}', \sigma_{12}', \sigma_{21}'$, and σ_{22}', by the relations

$$
\begin{aligned}
\sigma_{11}' &= \sigma_{11}\cos^2\theta + \sigma_{22}\sin^2\theta + (\sigma_{12} + \sigma_{21})\sin\theta\cos\theta, \\
\sigma_{12}' &= \sigma_{12}\cos^2\theta - \sigma_{21}\sin^2\theta - (\sigma_{11} - \sigma_{22})\sin\theta\cos\theta, \\
\sigma_{22}' &= \sigma_{11}\sin^2\theta + \sigma_{22}\cos^2\theta - (\sigma_{12} + \sigma_{21})\sin\theta\cos\theta, \\
\sigma_{21}' &= \sigma_{21}\cos^2\theta - \sigma_{12}\sin^2\theta - (\sigma_{11} - \sigma_{22})\sin\theta\cos\theta,
\end{aligned} \tag{2.83}
$$

where we have assumed a unit thickness for the triangular lamina and we obtain the last two relations from the first two by setting $\theta \to \theta + \pi/2$. This system can be written in terms of matrices as

$$
\begin{bmatrix} \sigma_{11}' & \sigma_{12}' \\ \sigma_{21}' & \sigma_{22}' \end{bmatrix} = \begin{bmatrix} \cos\theta & \sin\theta \\ -\sin\theta & \cos\theta \end{bmatrix} \begin{bmatrix} \sigma_{11} & \sigma_{12} \\ \sigma_{21} & \sigma_{22} \end{bmatrix} \begin{bmatrix} \cos\theta & -\sin\theta \\ \sin\theta & \cos\theta \end{bmatrix}, \tag{2.84}
$$

or

$$
\boldsymbol{\sigma}' = \boldsymbol{Q}\boldsymbol{\sigma}\,\boldsymbol{Q}^T. \tag{2.85}
$$

Using index notation, we have

$$
\sigma_{ij}' = Q_{im}\sigma_{mn}Q_{nj}^T = Q_{im}Q_{jn}\sigma_{mn}. \tag{2.86}
$$

Thus the stress components do transform in accordance with the transformation law for Cartesian tensors of rank 2. Using tensor notation, we have

$$
\boldsymbol{\sigma}' = \sigma_{ij}\boldsymbol{e}_i'\boldsymbol{e}_j', \quad \boldsymbol{\sigma} = \sigma_{ij}\boldsymbol{e}_i\boldsymbol{e}_j, \quad \boldsymbol{Q} = Q_{ij}\boldsymbol{e}_i\boldsymbol{e}_j, \tag{2.87}
$$

$$
\boldsymbol{\sigma}' = \boldsymbol{Q}\cdot\boldsymbol{\sigma}\cdot\boldsymbol{Q}^T. \tag{2.88}
$$

When the tensors involved are of the second rank or lower, we use matrix notation and write the preceding relation as

$$\boldsymbol{\sigma}' = \boldsymbol{Q}\boldsymbol{\sigma}\,\boldsymbol{Q}^T. \tag{2.89}$$

The following results are useful in dealing with tensor equations:

1. If all the components of a tensor vanish in one coordinate system, they vanish in all coordinate systems.
2. The sum and difference of two Cartesian tensors of rank r in N dimensions are also Cartesian tensors of rank r in N dimensions.

On the basis of the preceding results, the importance of tensor analysis may be summarized as follows: The form of an equation can have general validity with respect to any frame of reference only if every term in the equation has the same tensor characteristics.

2.6.2 Quotient Rule

Consider an array of N^3 quantities $A(i, j, k)$, with each index ranging over $1, 2, \ldots, N$. We do not know if it is a tensor. Suppose we know the product of $A(i, j, k)$ with an arbitrary tensor behaves as a tensor. Then we can use the **quotient rule** to establish the nature of $A(i, j, k)$. For example, let v_i be components of a vector, and if the product $A(i, j, k)v_i$ is known to yield a tensor of rank 2, B_{jk}, then

$$A(i, j, k)v_i = B_{jk}. \tag{2.90}$$

In a different coordinate system

$$A'(i, j, k)v_i' = B_{jk}'. \tag{2.91}$$

Using the transformation rule for the known tensors, we obtain

$$\begin{aligned} A'(i, j, k)v_i' &= Q_{jm}Q_{kn}B_{mn} \\ &= Q_{jm}Q_{kn}A(l, m, n)v_l \\ &= Q_{jm}Q_{kn}A(l, m, n)\,Q_{il}v_i', \end{aligned} \tag{2.92}$$

$$[A'(i, j, k) - Q_{jm}Q_{kn}Q_{il}A(l, m, n)]v_i' = 0. \tag{2.93}$$

Because v_i' are arbitrary, it follows that the quantity inside the brackets must be zero for all values of i. That is,

$$A'(i, j, k) = Q_{jm}Q_{kn}Q_{il}A(l, m, n). \tag{2.94}$$

Thus $\boldsymbol{A}$ is a third-rank tensor. This is known as the quotient rule.

2.6.3 Inner Products: Notation

Consider two second-rank tensors, $\boldsymbol{A}$ and $\boldsymbol{B}$. The product $A_{ij}B_{jk}$ can be expressed in tensor notation as

$$A_{ij}B_{jk}\boldsymbol{e}_i\boldsymbol{e}_k = \boldsymbol{A}\cdot\boldsymbol{B}. \tag{2.95}$$

In matrix notation, we have

$$A_{ij}B_{jk}\boldsymbol{e}_i\boldsymbol{e}_k = \boldsymbol{A}\boldsymbol{B}. \tag{2.96}$$

The matrix multiplication rule implements the inner product automatically. The product $A_{ij}B_{ji}$ that we obtain by letting $k = i$ can be written in the matrix notation as

$$A_{ij}B_{ji} = \mathrm{Tr}(\boldsymbol{A}\boldsymbol{B}), \tag{2.97}$$

and we use the **double-dot notation**

$$A_{ij}B_{ji} = \boldsymbol{A} : \boldsymbol{B} \tag{2.98}$$

in the tensor form.

2.7 Quadratic Forms and Eigenvalue Problems

Associated with a second-rank tensor (or matrix), a homogeneous quadratic form can be defined:

$$A = A_{ij}x_ix_j = \boldsymbol{x}^T\boldsymbol{A}\boldsymbol{x}. \tag{2.99}$$

If A_{ij} is unsymmetrical, when we construct the quadratic form the coefficient of x_ix_j comes out to $(A_{ij} + A_{ji})/2$. Thus we need to consider only symmetric matrices in connection with quadratic forms. In other words, if an unsymmetrical matrix A_{ij} is given, first we convert it to a symmetric matrix $A_{(ij)}$ when dealing with quadratic forms. The function A is called a homogeneous function of the second degree as it takes the value $\mu^2 A$ if all x_i are replaced with μx_i. We often encounter problems in which we have to find the extremum or stationary value of A with respect to x_i with the added normalization constraint $|\boldsymbol{x}| = 1$. We can perform constrained extremization by defining a modified function A^* using a Lagrange multiplier λ in the form

$$A^* = A - \lambda(x_ix_i - 1) \tag{2.100}$$

and setting its derivatives with respect to x_i to zero. This gives

$$\frac{\partial A^*}{\partial x_i} = 2\,(A_{ij}x_j - \lambda x_i) = 0. \tag{2.101}$$

This system of equations can be put in the form

$$\boldsymbol{A}\boldsymbol{x} = \lambda\boldsymbol{x} \quad \text{or} \quad [\boldsymbol{A} - \lambda\boldsymbol{I}]\boldsymbol{x} = \boldsymbol{0}. \tag{2.102}$$

This homogeneous system of simultaneous equations always has a trivial solution $\boldsymbol{x} = \boldsymbol{0}$ that does not satisfy our normalization constraint $|\boldsymbol{x}| = 1$. Recall that, for a nontrivial solution, we require

$$\det[\boldsymbol{A} - \lambda \boldsymbol{I}] = 0. \tag{2.103}$$

For a given matrix $\boldsymbol{A}$, this is possible for only special, discrete values of λ. Assuming we have a 3×3 matrix, when the preceding determinant is expanded, we get a cubic equation known as the characteristic equation of the matrix $\boldsymbol{A}$:

$$-\lambda^3 + I_{A1}\lambda^2 - I_{A2}\lambda + I_{A3} = 0, \tag{2.104}$$

where the coefficients I_{A1}, I_{A2}, and I_{A3} can be written as

$$I_{A1} = A_{ii}, \quad I_{A2} = \frac{1}{2}(A_{ii}A_{jj} - A_{ij}A_{ij}), \quad I_{A3} = \det \boldsymbol{A}. \tag{2.105}$$

These coefficients are also known as the three **invariants** of the matrix $\boldsymbol{A}$. If we denote the three roots of cubic equation (2.104) by λ_1, λ_2, and λ_3, we see

$$\begin{aligned} I_{A1} &= \lambda_1 + \lambda_2 + \lambda_3, \\ I_{A2} &= \lambda_1\lambda_2 + \lambda_2\lambda_3 + \lambda_3\lambda_1, \\ I_{A3} &= \lambda_1\lambda_2\lambda_3. \end{aligned} \tag{2.106}$$

The significance of the term invariant will be clear if we use a new coordinate system x_i' that we obtain by rotating the x_i system. Let

$$x_i' = Q_{ij}x_j \quad \text{or} \quad \boldsymbol{x} = \boldsymbol{Q}^T\boldsymbol{x}'. \tag{2.107}$$

Using this in the quadratic form, Eq. (2.99), we get

$$A = \boldsymbol{x}'^T\boldsymbol{Q}\boldsymbol{A}\boldsymbol{Q}^T\boldsymbol{x}' = \boldsymbol{x}'^T\boldsymbol{A}'\boldsymbol{x}', \tag{2.108}$$

where

$$\boldsymbol{A}' = \boldsymbol{Q}\boldsymbol{A}\boldsymbol{Q}^T \quad \text{or} \quad \boldsymbol{A} = \boldsymbol{Q}^T\boldsymbol{A}'\boldsymbol{Q}. \tag{2.109}$$

Substituting for $\boldsymbol{A}$ in Eq. (2.103), we get

$$\begin{aligned} \det[\boldsymbol{Q}^T\boldsymbol{A}'\boldsymbol{Q} - \lambda\boldsymbol{I}] &= 0, \\ \det[\boldsymbol{Q}^T\boldsymbol{A}'\boldsymbol{Q} - \lambda\boldsymbol{Q}^T\boldsymbol{I}\boldsymbol{Q}] &= 0, \\ \det[\boldsymbol{Q}^T(\boldsymbol{A}' - \lambda\boldsymbol{I})\boldsymbol{Q}] &= 0, \\ \det\boldsymbol{Q}^T\det(\boldsymbol{A}' - \lambda\boldsymbol{I})\det\boldsymbol{Q} &= 0, \\ \det(\boldsymbol{A}' - \lambda\boldsymbol{I}) &= 0, \end{aligned} \tag{2.110}$$

where we have used $\det \boldsymbol{Q} = \det \boldsymbol{Q}^T \neq 0$. From this, we have the characteristic equation

$$-\lambda^3 + I_{A'1}\lambda^2 - I_{A'2}\lambda + I_{A'3} = 0. \tag{2.111}$$

Comparing this equation with Eq. (2.104) and noting that the three roots of the cubic are the same, we clearly see that

$$I_{A1} = I_{A'1}, \quad I_{A2} = I_{A'2}, \quad I_{A3} = I_{A'3}. \tag{2.112}$$

Thus these three quantities remain invariant under coordinate rotation.

If we substitute $\lambda = \lambda_1$ in Eq. (2.102), the determinant of the system is zero, which implies that only two of the three equations are independent. We can solve the system to get a nonunique solution with one arbitrary parameter. The normalization condition $|\boldsymbol{x}| = 1$ determines the value of the parameter. The vector found this way is called an eigenvector of $\boldsymbol{A}$. It is known that, for a symmetric matrix, all eigenvalues are real and the eigenvectors corresponding to distinct eigenvalues are orthogonal. Further, even when the eigenvalues are not distinct, three mutually orthogonal eigenvectors can be constructed if the matrix is symmetric. These eigenvectors are called the principal directions of matrix $\boldsymbol{A}$, and the eigenvalues are the principal values. Corresponding to the three eigenvalues λ_i, the three eigenvectors are denoted by $\boldsymbol{x}^{(i)}$. An interesting but simple relation can be found between the extremum values of the quadratic form A and the Lagrange multipliers (eigenvalues) as follows: Multiply the equation

$$\boldsymbol{A}\boldsymbol{x}^{(1)} = \lambda_1 \boldsymbol{x}^{(1)} \tag{2.113}$$

by $\boldsymbol{x}^{(1)T}$ to get

$$A = \boldsymbol{x}^{(1)T} \boldsymbol{A}\boldsymbol{x}^{(1)} = \lambda_1 \boldsymbol{x}^{(1)T} \boldsymbol{x}^{(1)} = \lambda_1. \tag{2.114}$$

Thus the three eigenvalues are the extremum values of the quadratic form along the principal directions. If all the eigenvalues are positive, the quadratic form is called positive definite and the matrix is called a positive-definite matrix. In the 3D space, we can visualize an ellipsoid corresponding to the quadratic form, with its major axis, minor axis, and an "in-between" axis lying along the principal directions.

An interesting property of any square matrix is given by the Cayley–Hamilton theorem, which states that matrix $\boldsymbol{A}$ satisfies its own characteristic equation. For a 3×3 matrix, we have

$$-\boldsymbol{A}^3 + I_{A1}\boldsymbol{A}^2 - I_{A2}\boldsymbol{A} + I_{A3}\boldsymbol{I} = \boldsymbol{O}, \tag{2.115}$$

where the quantity on the right-hand side is a square null matrix. As a consequence, $\boldsymbol{A}^3$ can be expressed in terms of lower powers of $\boldsymbol{A}$, and when this step is used iteratively, any matrix polynomial can be reduced to a quadratic for the 3×3 case.

2.7.1 Example: Eigenvalue Problem

Obtain the eigenvalues and eigenvectors of the matrix

$$\boldsymbol{A} = \frac{1}{2} \begin{bmatrix} 7 & 1 & 0 \\ 1 & 7 & 0 \\ 0 & 0 & 4 \end{bmatrix}. \tag{2.116}$$

The characteristic equation is obtained from

$$\det(\boldsymbol{A} - \lambda \boldsymbol{I}) = 0, \tag{2.117}$$

as

$$[(7 - 2\lambda)^2 - 1][4 - 2\lambda] = 0, \quad \text{or} - \lambda^3 + 9\lambda^2 - 26\lambda + 24 = 0. \tag{2.118}$$

The invariants of the matrix are

$$I_{A1} = 9, \quad I_{A2} = 26, \quad I_{A3} = 24. \tag{2.119}$$

From the first of the equations in Eq. (2.118), we see the eigenvalues are

$$\lambda_1 = 4, \quad \lambda_2 = 3, \quad \lambda_3 = 2, \tag{2.120}$$

and the corresponding eigenvectors are

$$\boldsymbol{x}^{(1)} = \frac{1}{\sqrt{2}} \begin{Bmatrix} 1 \\ 1 \\ 0 \end{Bmatrix}, \quad \boldsymbol{x}^{(2)} = \frac{1}{\sqrt{2}} \begin{Bmatrix} -1 \\ 1 \\ 0 \end{Bmatrix}, \quad \boldsymbol{x}^{(3)} = \begin{Bmatrix} 0 \\ 0 \\ 1 \end{Bmatrix}. \tag{2.121}$$

2.7.2 Diagonalization and Polar Decomposition

Using the three eigenvectors as the three columns, we can construct the modal matrix of the given matrix $\boldsymbol{A}$. Let us denote the modal matrix by $\boldsymbol{M}$. Then,

$$\boldsymbol{M} = [\boldsymbol{x}^{(1)}, \boldsymbol{x}^{(2)}, \boldsymbol{x}^{(3)}]. \tag{2.122}$$

Then it is clear that

$$\boldsymbol{M}^T \boldsymbol{A} \boldsymbol{M} = \begin{bmatrix} \lambda_1 & 0 & 0 \\ 0 & \lambda_2 & 0 \\ 0 & 0 & \lambda_3 \end{bmatrix}. \tag{2.123}$$

This procedure amounts to choosing a new Cartesian system aligned with the principal directions of the quadratic. That is, we choose a principal coordinate system x'_i, using

$$x_i = M_{ij} x'_j \quad \text{or} \quad \boldsymbol{x}' = \boldsymbol{M}^T \boldsymbol{x}. \tag{2.124}$$

Multiplying Eq. (2.123) by $\boldsymbol{M}$ from the left and by $\boldsymbol{M}^T$ from the right, we can express matrix $\boldsymbol{A}$ as

$$\boldsymbol{A} = \sum_{i=1}^{3} \lambda_i \boldsymbol{x}^{(i)} \boldsymbol{x}^{(i)T}. \tag{2.125}$$

This is called the spectral representation of $\boldsymbol{A}$.

So far we have been concerned with a symmetrical matrix $\boldsymbol{A}$. The polar decomposition applies to any square matrix $\boldsymbol{B}$. Polar decomposition refers to factoring $\boldsymbol{B}$ in the form

$$\boldsymbol{B} = \boldsymbol{R}\boldsymbol{U} \quad \text{or} \quad \boldsymbol{B} = \boldsymbol{V}\boldsymbol{R}, \tag{2.126}$$

where $\boldsymbol{U}$ and $\boldsymbol{V}$ are symmetric matrices and $\boldsymbol{R}$ is an orthogonal (rotation) matrix.

Using $\boldsymbol{R}\boldsymbol{R}^T = \boldsymbol{I}$, we get

$$\boldsymbol{U}^2 = \boldsymbol{B}^T\boldsymbol{B} \quad \text{and} \quad \boldsymbol{V}^2 = \boldsymbol{B}\boldsymbol{B}^T. \tag{2.127}$$

To find $\boldsymbol{U}$ and $\boldsymbol{V}$, we need the square roots of the matrices on the right-hand sides of the preceding equations. After $\boldsymbol{U}$ or $\boldsymbol{V}$ is found, $\boldsymbol{R}$ can be expressed as

$$\boldsymbol{R} = \boldsymbol{B}\boldsymbol{U}^{-1} = \boldsymbol{V}^{-1}\boldsymbol{B}. \tag{2.128}$$

This brings us to finding the square root of a symmetric matrix, say $\boldsymbol{C}$ or, in general, any function $\boldsymbol{F}[\boldsymbol{C}]$ of a matrix. We begin by assuming the function $\boldsymbol{F}[\boldsymbol{C}]$ has a converging infinite series expansion in $\boldsymbol{C}$:

$$\boldsymbol{F}[\boldsymbol{C}] = \sum_0 a_i \boldsymbol{C}^i. \tag{2.129}$$

The functions we have in mind are $\boldsymbol{C}^{1/2}$, $\sin[\boldsymbol{C}]$, $\exp[\boldsymbol{C}]$, etc. The corresponding functions of a single variable, say x, are $x^{1/2}$, $\sin x$, $\exp x$, etc. The generic matrix function $\boldsymbol{F}$ has the corresponding function of a single variable F. Our symmetric matrix $\boldsymbol{C}$ has three orthogonal eigenvectors $\boldsymbol{x}^{(i)}$ with the corresponding eigenvalues λ_i. Multiplying Eq. (2.129) from the right by $\boldsymbol{x}^{(i)}$, we see that $F(\lambda_i)$ is the eigenvalue of $\boldsymbol{F}$ corresponding to the eigenvector, $\boldsymbol{x}^{(i)}$. The eigenvalues of $[\boldsymbol{C}]^{1/2}$, $\sin[\boldsymbol{C}]$, and $\exp[\boldsymbol{C}]$ are $\lambda_i^{1/2}$, $\sin\lambda_i$, and $\exp\lambda_i$, respectively. Next, we use the Cayley–Hamilton theorem to reduce Eq. (2.129) to a quadratic in $\boldsymbol{C}$ (provided we are still dealing with 3×3 matrices):

$$\boldsymbol{F}[\boldsymbol{C}] = c_0\boldsymbol{I} + c_1\boldsymbol{C} + c_2\boldsymbol{C}^2. \tag{2.130}$$

Then the eigenvalues satisfy

$$F(\lambda_i) = c_0 + c_1\lambda_i + c_2\lambda_i^2, \quad i = 1, 2, 3. \tag{2.131}$$

For example, to find the square root of $\boldsymbol{C}$, we use

$$\lambda_i^{1/2} = c_0 + c_1\lambda_i + c_2\lambda_i^2, \quad i = 1, 2, 3, \tag{2.132}$$

and solve for c_i. We substitute these coefficients in Eq. (2.130) to get $\boldsymbol{C}^{1/2}$. We may anticipate complications when the eigenvalues are not distinct. If $\lambda_2 = \lambda_1$, we follow the Frobenius limiting process. We begin by assuming $\lambda_2 = \lambda_1 + \epsilon$. Then, one of the three equations for c_i becomes

$$F(\lambda_1 + \epsilon) = c_0 + c_1(\lambda_1 + \epsilon) + c_2(\lambda_1 + \epsilon)^2. \tag{2.133}$$

If we subtract the equation corresponding to λ_1 and divide by ϵ, we have

$$\frac{F(\lambda_1 + \epsilon) - F(\lambda_1)}{\epsilon} = c_1 + c_2[(\lambda_1 + \epsilon)^2 - \lambda_1^2]/\epsilon. \tag{2.134}$$

As $\epsilon \to 0$, we obtain the needed additional equation,

$$F'(\lambda_1) = c_1 + 2c_2\lambda_1. \tag{2.135}$$

All of our discussions in the last two subsections were limited to 3×3 matrices. Extension of the preceding results for square matrices of any order is straightforward.

2.7.3 Example: Polar Decomposition

Suppose we want to factor the matrix

$$\boldsymbol{B} = \frac{1}{5}\begin{bmatrix} 17 & -11 & 0 \\ 19 & 23 & 0 \\ 0 & 0 & 15 \end{bmatrix}. \tag{2.136}$$

The factors $\boldsymbol{U}$, $\boldsymbol{V}$, and $\boldsymbol{R}$ are obtained from

$$\boldsymbol{U}^2 = \boldsymbol{C} \equiv \boldsymbol{B}^T\boldsymbol{B} = \begin{bmatrix} 26 & 10 & 0 \\ 10 & 26 & 0 \\ 0 & 0 & 9 \end{bmatrix}, \quad \boldsymbol{R} = \boldsymbol{B}\boldsymbol{U}^{-1}, \quad \boldsymbol{V} = \boldsymbol{B}\boldsymbol{R}^T. \tag{2.137}$$

The eigenvalues of $\boldsymbol{C}$ are

$$\lambda_1 = 9, \quad \lambda_2 = 16, \quad \lambda_3 = 36.$$

Using the expansion

$$\sqrt{\lambda_i} = c_0 + c_1\lambda_i + c_2\lambda_i^2,$$

we have the simultaneous equations

$$3 = c_0 + 9c_1 + 9^2c_2, \quad 4 = c_0 + 16c_1 + 16^2c_2, \quad 6 = c_0 + 36c_1 + 36^2c_2.$$

The solutions of this system are

$$c_0 = 52/35, \quad c_1 = 23/126, \quad c_2 = -1/630.$$

Writing

$$\boldsymbol{C}^{1/2} = c_0\boldsymbol{I} + c_1\boldsymbol{C} + c_2\boldsymbol{C}^2,$$

we get

$$\boldsymbol{U} = \boldsymbol{C}^{1/2} = \begin{bmatrix} 5 & 1 & 0 \\ 1 & 5 & 0 \\ 0 & 0 & 3 \end{bmatrix},$$

$$\boldsymbol{R} = \frac{1}{5}\begin{bmatrix} 4 & 3 & 0 \\ -3 & 4 & 0 \\ 0 & 0 & 5 \end{bmatrix},$$

$$\boldsymbol{V} = \frac{1}{25}\begin{bmatrix} 101 & 7 & 0 \\ 7 & 149 & 0 \\ 0 & 0 & 75 \end{bmatrix}.$$

SUGGESTED READING

Fung, Y. C. (1993). *First Course in Continuum Mechanics* 3rd ed., Prentice-Hall.
Jeffreys, H. (1931). *Cartesian Tensors*, Cambridge University Press.
Knowles, J. K. (1998). *Linear Vector Spaces and Cartesian Tensors*, Oxford University Press.
Temple, G. (1960). *Cartesian Tensors: An Introduction*, Dover.

EXERCISES

Use index notation and the summation convention wherever necessary in the following problems.

2.1. If $\boldsymbol{v}$ is a vector and $\boldsymbol{n}$ is a unit vector, show that

$$\boldsymbol{v} = (\boldsymbol{n} \cdot \boldsymbol{v})\boldsymbol{n} + (\boldsymbol{n} \times \boldsymbol{v}) \times \boldsymbol{n}.$$

2.2. With the help of the e–δ identity, show that any three vectors $\boldsymbol{a}$, $\boldsymbol{b}$, and $\boldsymbol{c}$ satisfy

$$\boldsymbol{a} \times (\boldsymbol{b} \times \boldsymbol{c}) = (\boldsymbol{a} \cdot \boldsymbol{c})\boldsymbol{b} - (\boldsymbol{a} \cdot \boldsymbol{b})\boldsymbol{c}.$$

Verify your result by using the method of determinants for the cross products.

2.3. Complete the following identities by obtaining the missing part $\boldsymbol{X}$.

(a) $\nabla\cdot(\boldsymbol{u} \times \boldsymbol{v}) = \boldsymbol{v} \cdot \nabla \times \boldsymbol{u} + \boldsymbol{X}$
(b) $\nabla(\boldsymbol{u} \cdot \boldsymbol{v}) = (\boldsymbol{u} \cdot \nabla\boldsymbol{v}) + \boldsymbol{X}$
(c) $\nabla \times (\boldsymbol{u} \times \boldsymbol{v}) = \boldsymbol{u}\nabla \cdot \boldsymbol{v} + \boldsymbol{X}$

2.4. If the three functions u, v, and w of x_1, x_2, and x_3 satisfy

$$F(u, v, w) = 0,$$

show that

$$\nabla u \cdot \nabla v \times \nabla w = 0.$$

2.5. Prove that all the components of a tensor vanish in all Cartesian coordinate systems if they vanish in one coordinate system.

2.6. Prove that the sum (and difference) of two tensors of rank r in an N-dimensional space is also a tensor of rank r.

2.7. For any four vectors $\boldsymbol{s}$, $\boldsymbol{t}$, $\boldsymbol{u}$, and $\boldsymbol{v}$, show that

$$(\boldsymbol{s} \times \boldsymbol{t}) \cdot (\boldsymbol{u} \times \boldsymbol{v}) = (\boldsymbol{s} \cdot \boldsymbol{u})(\boldsymbol{t} \cdot \boldsymbol{v}) - (\boldsymbol{s} \cdot \boldsymbol{v})(\boldsymbol{t} \cdot \boldsymbol{u}).$$

2.8. Show that

$$\nabla \cdot (\boldsymbol{u} \times \boldsymbol{v}) = \boldsymbol{v} \cdot \nabla \times \boldsymbol{u} - \boldsymbol{u} \cdot \nabla \times \boldsymbol{v}.$$

2.9. For a twice-differentiable, vector-valued function $\boldsymbol{v}$, show that

$$\nabla \times (\nabla \times \boldsymbol{v}) = \nabla(\nabla \cdot \boldsymbol{v}) - (\nabla \cdot \nabla)\boldsymbol{v}.$$

2.10. If $\boldsymbol{r} = \boldsymbol{e}_i x_i$ and $r^2 = x_i x_i$, show that (Fung, 1993)

(a) $\nabla \cdot (r^n \boldsymbol{r}) = (n + 3)r^n$,
(b) $\nabla \times (r^n \boldsymbol{r}) = 0$,

(c) $\nabla \cdot \nabla r^n = n(n+1)r^{(n-2)}$.
(d) If F is any differentiable function, show that

$$\nabla \times [F(r)\mathbf{r}] = \mathbf{0}.$$

2.11. Show that the differential equation

$$(\nabla^2 + k^2)\psi = 0,$$

in three dimensions, admits the solution

$$\psi = \frac{e^{\pm ikr}}{r}.$$

2.12. If $\boldsymbol{A}$ and $\boldsymbol{B}$ are tensors of rank m and n, respectively, in N dimensions, show that

(a) the rank of $\boldsymbol{A} \cdot \boldsymbol{B}$ is $m + n - 2$,
(b) the rank of $\boldsymbol{A} \times \boldsymbol{B}$ is $m + n - 1$, and
(c) the rank of $\boldsymbol{A} \bigotimes \boldsymbol{B}$ is $m + n$.

2.13. If $\boldsymbol{T}$ is a two-dimensional, second-rank, unsymmetrical tensor and $\boldsymbol{n}$ is a two-dimensional vector (two-vector), compare $\boldsymbol{n} \cdot \boldsymbol{T} \cdot \boldsymbol{n}$, $(\boldsymbol{T} \cdot \boldsymbol{n}) \cdot \boldsymbol{n}$, and $\boldsymbol{n} \cdot (\boldsymbol{n} \cdot \boldsymbol{T})$.

2.14. Consider two Cartesian coordinate systems in two dimensions, x_i' and x_i, with the x_1' axis rotated by angle θ in the counterclockwise direction from the x_1 axis. For a position vector $\boldsymbol{r}$, which makes an angle α with the x_1 axis, obtain the components in the two systems and relate them as

$$\boldsymbol{x}' = \boldsymbol{Q}\boldsymbol{x}.$$

Consider a new position vector $\boldsymbol{s}$ obtained by rotating $\boldsymbol{r}$ by an angle, θ. Obtain the components of $\boldsymbol{s}$ in the x-coordinate system:

$$\boldsymbol{s} = \boldsymbol{P}\boldsymbol{x}.$$

Compare the two rotation matrices $\boldsymbol{Q}$ and $\boldsymbol{P}$.

2.15. Compute the eigenvalues and the normalized eigenvectors of the matrix

$$\boldsymbol{C} = \begin{bmatrix} 4 & 2 & 0 \\ 2 & 4 & 0 \\ 0 & 0 & 1 \end{bmatrix}.$$

Construct the modal matrix $\boldsymbol{M}$ and show that $\boldsymbol{M}^T\boldsymbol{C}\boldsymbol{M}$ is a diagonal matrix $\boldsymbol{D}$, with the eigenvalues along the diagonal. Obtain the square root of this diagonal matrix $\boldsymbol{D}^{1/2}$. Show that $\boldsymbol{M}\boldsymbol{D}^{1/2}\boldsymbol{M}^T$ is the square root of $\boldsymbol{C}$.

2.16. Obtain the factors $\boldsymbol{U}$, $\boldsymbol{V}$, and $\boldsymbol{R}$ in the polar decomposition of the matrix

$$\boldsymbol{B} = \begin{bmatrix} \sqrt{3} & 1 & 0 \\ 0 & 2 & 0 \\ 0 & 0 & 1 \end{bmatrix}$$

in the form

$$\boldsymbol{B} = \boldsymbol{RU} = \boldsymbol{VR}.$$

2.17. For the matrix

$$\boldsymbol{B} = \begin{bmatrix} 0 & 1 \\ -1 & 0 \end{bmatrix},$$

show that

$$e^{\boldsymbol{B}\theta} = \boldsymbol{I}\cos\theta + \boldsymbol{B}\sin\theta,$$

where θ is a real number.

2.18. For

$$\boldsymbol{C} = \begin{bmatrix} 5 & 4 \\ 0 & 5 \end{bmatrix},$$

compute $\boldsymbol{C}^{10}$ and $\sin \boldsymbol{C}$.

2.19. In the case of an unsymmetric matrix $\boldsymbol{A}$, show that $\boldsymbol{A}$ and $\boldsymbol{A}^T$ have the same eigenvalues, $\lambda^{(i)}$, $i = 1, 2, 3$, and if $\boldsymbol{x}^{(i)}$ are three independent eigenvectors of $\boldsymbol{A}$ and $\boldsymbol{y}^i$, $i = 1, 2, 3$ are the eigenvectors of $\boldsymbol{A}^T$ (these are also called the left eigenvectors of $\boldsymbol{A}$), show that the two sets of eigenvectors can be selected to have

$$\boldsymbol{x}^{(i)T}\boldsymbol{y}^{(j)} = \delta_{ij}.$$

This relation illustrates **biorthogonality** of two families of vectors.

2.20. Using the notation of the preceding problem, show that $\boldsymbol{A}$ has the **spectral representation**

$$\mathbf{A} = \sum_{i=1}^{3} \lambda^{(i)}\boldsymbol{x}^{(i)}\mathbf{y}^{(i)T},$$

when there are three independent eigenvectors $\boldsymbol{x}^{(i)}$.

2.21. Obtain the eigenvalues and the left and right eigenvectors of the matrix

$$\mathbf{A} = \begin{bmatrix} 1 & -1 & 1 \\ -1 & 1 & -1 \\ -1 & 1 & -1 \end{bmatrix}.$$

What is the spectral representation for this matrix?

3 General Tensors

There are a number of continuum mechanics books in the literature, written with the general tensor formulation. An essential characteristic of general tensors is the use of curvilinear coordinates. We may imagine a Cartesian grid placed inside a continuum. As the continuum deforms, the grid lines deform with it into a mesh of curves. Thus general tensor formulations are convenient to describe continuum mechanics. Although this chapter is not required to follow our presentation of continuum mechanics in terms of Cartesian tensors, it will be of value to extend the students' knowledge through additional reading. In modern computational mechanics, curvilinear grids are often used and the governing equations in general tensor formulation are needed. As an added bonus, students will be able to follow the theory of shells and the general theory of relativity!

In the previous chapter we saw systems of Cartesian coordinates and transformations of tensors between two coordinates that are related linearly. When we have to deal with curvilinear coordinates, the coordinate transformations are in general nonlinear and the coordinates may not form an orthogonal system. Representation of tensors in such a system depends on the directions of local tangents and normals to the coordinate surfaces. It is conventional to use superscripts to denote the coordinate curves. The reason for this will be clarified later. Let us begin with a Cartesian system with labels x^1, x^2, and x^3, transforming into curvilinear system ξ^1, ξ^2, and ξ^3. A two-dimensional (2D) illustration of this is shown in Fig. 3.1. We assume that the relation between x^i and ξ^j,

$$x^i = x^i(\xi^1, \xi^2, \xi^3), \tag{3.1}$$

and its inverse,

$$\xi^j = \xi^j(x^1, x^2, x^3), \tag{3.2}$$

are locally one-to-one. What is meant by "locally one-to-one" is that in the neighborhood of a point, (ξ^1, ξ^2, ξ^3), the increments dx^i of the coordinates x^i are linearly related to the increments $d\xi^j$ of ξ^j, and these relations can be inverted.

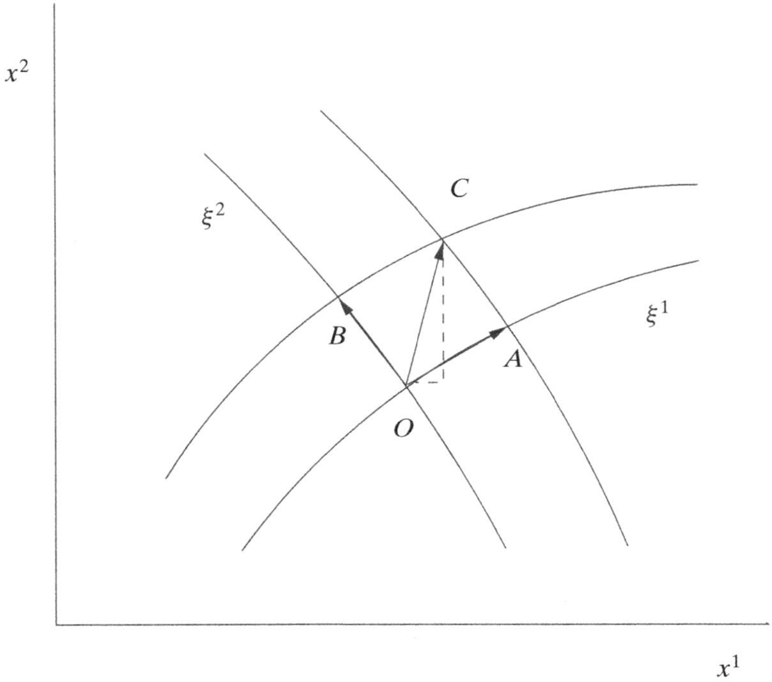

Figure 3.1. Distance element in a curvilinear system.

We have

$$dx^i = \frac{\partial x^i}{\partial \xi^j} d\xi^j \tag{3.3}$$

and the matrix

$$\boldsymbol{L} = [L_{ji}] = \left[\frac{\partial x^i}{\partial \xi^j}\right], \tag{3.4}$$

which is evaluated at (ξ^1, ξ^2, ξ^3), is nonsingular. We use the notation

$$L = \det[L_{ji}]. \tag{3.5}$$

Referring to Fig. 3.1, the distance vector $\overrightarrow{OC}$ can be expressed in terms of the increments in the Cartesian coordinates dx^i, as well as in terms of the increments in the curvilinear coordinates $d\xi^i$.

With

$$\overrightarrow{OA} = \boldsymbol{g}_1 d\xi^1, \quad \overrightarrow{OB} = \boldsymbol{g}_2 d\xi^2, \tag{3.6}$$

we have the vector sum,

$$\overrightarrow{OC} = dx^i \boldsymbol{e}_i = \boldsymbol{g}_i d\xi^i. \tag{3.7}$$

From this, the base vector, $\boldsymbol{g}_i$ can be related to the Cartesian base vectors $\boldsymbol{e}_i$ as

$$\boldsymbol{g}_i = \frac{\partial x^j}{\partial \xi^i} \boldsymbol{e}_j = L_{ij} \boldsymbol{e}_j. \tag{3.8}$$

A base vector $\boldsymbol{g}_i$ is tangential to the curve ξ^i. Here the "curve ξ^1" refers to the curve along which ξ^1 changes while ξ^2 and ξ^3 remain constant. Unlike Cartesian coordinates, curvilinear coordinates may not have the dimension of length. As an example, we use angles as coordinates in polar and spherical systems. The term $\boldsymbol{g}_i d\xi^i$ must have the dimension of length, and if $d\xi^i$ is nondimensional, $\boldsymbol{g}_i$ has the dimension of length. In general, then, all the $\boldsymbol{g}_i$ may not have the same dimensions and they are not unit vectors.

The distance between O and C can be expressed in the form

$$|OC|^2 = dx^k dx^k = \boldsymbol{g}_i \cdot \boldsymbol{g}_j d\xi^i d\xi^j = g_{ij} d\xi^i d\xi^j, \tag{3.9}$$

where we have defined

$$g_{ij} = \boldsymbol{g}_i \cdot \boldsymbol{g}_j = \frac{\partial x^k}{\partial \xi^i}\frac{\partial x^k}{\partial \xi^j}, \quad [g] = \boldsymbol{L}\boldsymbol{L}^T. \tag{3.10}$$

The elements g_{ij} are called the components of the Riemann metric tensor $\boldsymbol{g}$ associated with the curvilinear system. In the distance formula, when a Cartesian system is used we have the Euclidean distance measure $dx^k dx^k$, and when a curvilinear system is used we have the Riemann measure $g_{ij} d\xi^i d\xi^j$.

As mentioned earlier, $\boldsymbol{g}_1$ is tangent to the coordinate curve ξ^1, and $\boldsymbol{g}_1$ and $\boldsymbol{g}_2$ lie in the tangent plane of the surface formed by ξ^1 and ξ^2 curves. We can construct a normal $\boldsymbol{g}^3$ to this surface with its length adjusted to have $\boldsymbol{g}^3 \cdot \boldsymbol{g}_3 = 1$. Completing this scheme, we obtain three new vectors $\boldsymbol{g}^i$ with the properties

$$\boldsymbol{g}_i \cdot \boldsymbol{g}^j = \delta^i_j, \tag{3.11}$$

which is the Kronecker delta in the general tensor setting. In expanded form,

$$\begin{aligned}
&\boldsymbol{g}_1 \cdot \boldsymbol{g}^1 = 1, \qquad \boldsymbol{g}_1 \cdot \boldsymbol{g}^2 = 0, \qquad \boldsymbol{g}_1 \cdot \boldsymbol{g}^3 = 0,\\
&\boldsymbol{g}_2 \cdot \boldsymbol{g}^1 = 0, \qquad \boldsymbol{g}_2 \cdot \boldsymbol{g}^2 = 1, \qquad \boldsymbol{g}_2 \cdot \boldsymbol{g}^3 = 0,\\
&\boldsymbol{g}_3 \cdot \boldsymbol{g}^1 = 0, \qquad \boldsymbol{g}_3 \cdot \boldsymbol{g}^2 = 0, \qquad \boldsymbol{g}_3 \cdot \boldsymbol{g}^3 = 1.
\end{aligned} \tag{3.12}$$

We call $\boldsymbol{g}_i$ the covariant base vectors or simply base vectors and $\boldsymbol{g}^i$ the contravariant base vectors or reciprocal base vectors (see Fig. 3.2). Because $\boldsymbol{g}^i$ form normal vectors to the tangent planes, we have

$$\boldsymbol{g}^1 = k\boldsymbol{g}_2 \times \boldsymbol{g}_3, \tag{3.13}$$

where the constant of proportionality k is found from

$$\boldsymbol{g}_1 \cdot \boldsymbol{g}^1 = 1 = k\boldsymbol{g}_1 \cdot \boldsymbol{g}_2 \times \boldsymbol{g}_3. \tag{3.14}$$

As the **scalar product** or **triple product** of three vectors is invariant with respect to the permutations of the vectors, we find that the constant k, given by

$$k = (\boldsymbol{g}_1 \cdot \boldsymbol{g}_2 \times \boldsymbol{g}_3)^{-1}, \tag{3.15}$$

is invariant if we permute the integers in Eq. (3.13).

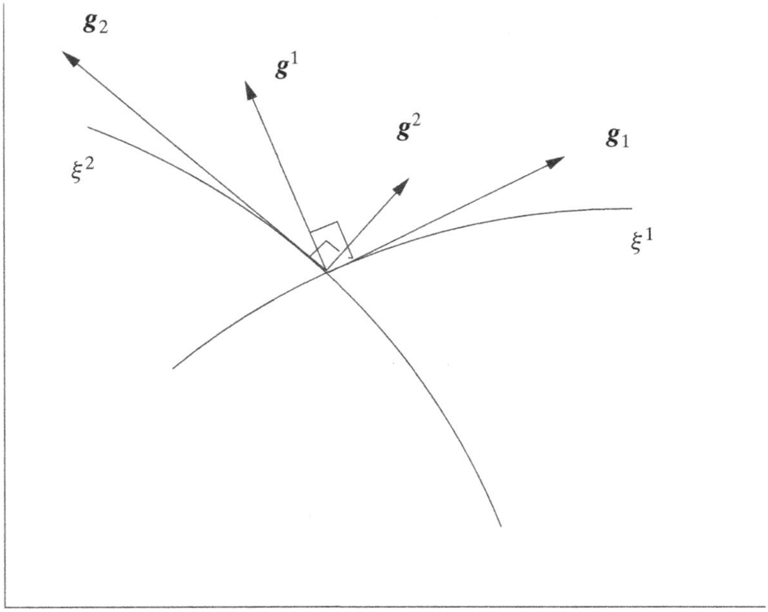

Figure 3.2. Base vectors and reciprocal base vectors.

Using the Cartesian representation of $\boldsymbol{g}_i$ in Eq. (3.8), we find that the scalar product is the determinant of $\boldsymbol{L}$, i.e.,

$$k = 1/L. \tag{3.16}$$

Because

$$\boldsymbol{g}_i = L_{ij}\boldsymbol{e}_j, \tag{3.17}$$

we have

$$g = \det[g_{ij}] = \det[\boldsymbol{g}_i \cdot \boldsymbol{g}_j] = |L_{ik}L_{jk}| = L^2. \tag{3.18}$$

Then

$$k = \frac{1}{L} = \frac{1}{\sqrt{g}}, \tag{3.19}$$

and

$$\boldsymbol{g}^1 = \frac{1}{\sqrt{g}}\boldsymbol{g}_2 \times \boldsymbol{g}_3, \quad \boldsymbol{g}^2 = \frac{1}{\sqrt{g}}\boldsymbol{g}_3 \times \boldsymbol{g}_1, \quad \boldsymbol{g}^3 = \frac{1}{\sqrt{g}}\boldsymbol{g}_1 \times \boldsymbol{g}_2. \tag{3.20}$$

Using the permutation symbol, we can write the preceding result as

$$\frac{1}{\sqrt{g}}\boldsymbol{g}_i \times \boldsymbol{g}_j = e_{ijk}\boldsymbol{g}^k. \tag{3.21}$$

As $\boldsymbol{g}^i$ is a vector in the three dimension (3D) space, we can express it as a linear combination of the base vectors $\boldsymbol{g}_j$,

$$\boldsymbol{g}^i = g^{ij}\boldsymbol{g}_j, \tag{3.22}$$

where we find the coefficient g^{ij} by using

$$\boldsymbol{g}^i \cdot \boldsymbol{g}_j = \delta^i_j = g^{ik} g_{kj}. \tag{3.23}$$

Thus the matrix $[g^{ij}]$ is the inverse of the matrix $[g_{ij}]$. That is

$$g_{ij} g^{jk} = \delta^k_i. \tag{3.24}$$

Inverting the system, Eq. (3.22), we get,

$$\boldsymbol{g}_i = g_{ij} \boldsymbol{g}^j. \tag{3.25}$$

The array g^{ij} is called the contravariant or reciprocal metric tensor, and $\boldsymbol{g}^i$ are also called the contravariant or reciprocal base vectors. We later verify that what we call tensors indeed follow the transformation law for tensors.

From the Cartesian permutation symbol, we construct the permutation tensors:

$$\epsilon_{ijk} = \sqrt{g} e_{ijk}, \quad \epsilon^{ijk} = \frac{1}{\sqrt{g}} e^{ijk}, \tag{3.26}$$

where we have written the permutation symbol as e^{ijk} to match the indices with those on the left-hand side.

With these we get

$$\boldsymbol{g}_i \times \boldsymbol{g}_j = \epsilon_{ijk} \boldsymbol{g}^k, \quad \boldsymbol{g}^k = \frac{1}{2} \epsilon^{ijk} \boldsymbol{g}_i \times \boldsymbol{g}_j. \tag{3.27}$$

3.1 Vectors and Tensors

As we have two systems of base vectors, $\boldsymbol{g}_i$ and $\boldsymbol{g}^i$, any vector $\boldsymbol{v}$ can be written as

$$\boldsymbol{v} = v_i \boldsymbol{g}^i = v^j \boldsymbol{g}_j, \tag{3.28}$$

where v_i are called the covariant components and v^j are called the contravariant components. The raising and lowering of indices, that is, the conversion of covariant to contravariant or vice versa can be performed by scalar multiplication of Eq. (3.28) by $\boldsymbol{g}_k$ or by $\boldsymbol{g}^k$;

$$\begin{aligned} v_i \boldsymbol{g}^i \cdot \boldsymbol{g}_k &= v^j \boldsymbol{g}_j \cdot \boldsymbol{g}_k, \\ v_i \delta^i_k &= v^j g_{jk}, \\ v_i &= g_{ij} v^j, \quad \text{or its inverse,} \quad v^i = g^{ij} v_j. \end{aligned} \tag{3.29}$$

For a second-rank tensor $\boldsymbol{A}$, we can have multiple representations:

$$\boldsymbol{A} = A_{ij} \boldsymbol{g}^i \otimes \boldsymbol{g}^j = A_i{}^j \boldsymbol{g}^i \otimes \boldsymbol{g}_j = A^i{}_j \boldsymbol{g}_i \otimes \boldsymbol{g}^j = A^{ij} \boldsymbol{g}_i \otimes \boldsymbol{g}_j. \tag{3.30}$$

We may relate the covariant, contravariant, and mixed components by using the metric tensor and its reciprocal in the form

$$\begin{aligned} A_{ij} &= g_{ik} g_{jl} A^{kl}, \\ A^{ij} &= g^{ik} g^{jl} A_{kl}, \\ A_i{}^j &= g_{ik} g^{jl} A^k{}_l. \end{aligned} \tag{3.31}$$

It has to be noted that in mixed components the horizontal positions of the superscript and subscript are important. If A_{ij} is symmetric in i and j, we can see that

$$A_i{}^j = A^j{}_i = A_i^j. \tag{3.32}$$

Once we have learned the basic operations using the metric tensor and its reciprocal, we are ready to define covariant, contravariant, and mixed arrays properly. For this we introduce two curvilinear systems, x^i and $\bar{x}^i$. The system x^i, being curvilinear, is a temporary notation; we will make it Cartesian again afterward. If an array $v(i)$ in the x^i system transforms into $\bar{v}(i)$ according to the rule

$$\bar{v}(i) = v(j)\frac{\partial x^j}{\partial \bar{x}^i}, \tag{3.33}$$

we call it a covariant array and use a subscript to write

$$\bar{v}_i = v_j \frac{\partial x^j}{\partial \bar{x}^i}. \tag{3.34}$$

It is a contravariant array if

$$\bar{v}(i) = v(j)\frac{\partial \bar{x}^j}{\partial x^i}, \tag{3.35}$$

and we write it with a superscript as

$$\bar{v}^i = v^j \frac{\partial \bar{x}^i}{\partial x^j}. \tag{3.36}$$

Note that the coordinate differentials themselves transform as

$$d\bar{x}^i = dx^j \frac{\partial \bar{x}^i}{\partial x^j}, \tag{3.37}$$

which is the rule for contravariant arrays: This is the reason for using superscripts for coordinates.

If we consider the partial derivatives of a scalar function

$$\bar{\phi}(\bar{x}^1, \bar{x}^2, \bar{x}^3) = \phi(x^1, x^2, x^3), \tag{3.38}$$

we find that they follow the covariant rule,

$$\frac{\partial \bar{\phi}}{\partial \bar{x}^i} = \frac{\partial \phi}{\partial x^j}\frac{\partial x^j}{\partial \bar{x}^i}. \tag{3.39}$$

As we discussed in the last chapter, geometry or physical considerations determine the tensor nature of a given array.

We may extend the preceding rules to second- or higher-rank tensors. For example,

$$\begin{aligned}
\text{covariant:} \quad & \bar{A}_{ij} = A_{mn}\frac{\partial x^m}{\partial \bar{x}^i}\frac{\partial x^n}{\partial \bar{x}^j}, \\
\text{contravariant:} \quad & \bar{A}^{ij} = A^{mn}\frac{\partial \bar{x}^i}{\partial x^m}\frac{\partial \bar{x}^j}{\partial x^n}, \\
\text{mixed:} \quad & \bar{A}^i{}_j = A^m{}_n\frac{\partial \bar{x}^i}{\partial x^m}\frac{\partial x^n}{\partial \bar{x}^j}.
\end{aligned} \tag{3.40}$$

In Cartesian coordinates, the base vectors $\boldsymbol{e}_i$ can be used to construct reciprocal base vectors $\boldsymbol{e}^i$. Because of their orthonormality, we end up with

$$\boldsymbol{e}^i = \boldsymbol{e}_i. \tag{3.41}$$

Thus subscripted quantities and the corresponding superscripted quantities are identical; or, there is no need to distinguish covariant components from contravariant components. We have already used this idea when we referred to δ^i_j and e^{ijk}.

Another convention consistently used in general tensor analysis is that the summation convention applies when the repeated pair of indices occurs in subscript–superscript form. That is, we should write the distance formula as

$$(ds)^2 = dx^k dx_k = g_{ij} d\xi^i d\xi^j. \tag{3.42}$$

3.2 Physical Components

As mentioned earlier, in general tensor analysis the base vectors may not be unit vectors, nor may they have the same dimensions. However, we may convert them to unit vectors by dividing them by their magnitude. For example, the relation

$$\boldsymbol{v} = v^i \boldsymbol{g}_i \tag{3.43}$$

can be written as

$$\boldsymbol{v} = v^1 \sqrt{g_{11}} \frac{\boldsymbol{g}_1}{\sqrt{g_{11}}} + \cdots + . \tag{3.44}$$

From this we can extract the physical component:

$$v^{(1)} = v^1 \sqrt{g_{11}}. \tag{3.45}$$

Similarly, for the covariant components in

$$\boldsymbol{v} = v_i \boldsymbol{g}^i, \tag{3.46}$$

the physical components are of the form

$$v_{(1)} = v_1 \sqrt{g^{11}}. \tag{3.47}$$

3.3 Tensor Calculus

Tensor calculus deals with differentiating and integrating tensors in curvilinear coordinates. In the Cartesian system the base vectors are constants, and

$$\frac{\partial \boldsymbol{u}}{\partial x^k} = \boldsymbol{e}_i \partial_k u^i. \tag{3.48}$$

In the case of general tensors in a curvilinear system, ξ, the base vectors are not constants, and

$$\frac{\partial \boldsymbol{u}}{\partial \xi^k} = \boldsymbol{g}_i \partial_k u^i + u^i \partial_k \boldsymbol{g}_i. \tag{3.49}$$

The derivatives of the base vectors in the second term can be written as a linear combination of the three base vectors, to have

$$\partial_k \boldsymbol{g}_i = \Gamma^j_{ki} \boldsymbol{g}_j, \tag{3.50}$$

where the coefficients Γ^j_{ki} are called the Christoffel symbols of the second kind. With this, the derivative of a vector can be written as

$$\frac{\partial \boldsymbol{u}}{\partial \xi^k} = (\partial_k u^i + u^j \Gamma^i_{jk}) \boldsymbol{g}_i. \tag{3.51}$$

We define the covariant derivative of a contravariant component u^i as

$$D_k u^i = \partial_k u^i + \Gamma^i_{jk} u^j. \tag{3.52}$$

Similarly, the covariant derivative of a covariant component becomes

$$D_k u_i = \partial_k u_i - \Gamma^j_{ik} u_j. \tag{3.53}$$

The symbol D_k has different meanings, depending on the tensor components on which it operates.

The Christoffel symbols of the first kind are defined as

$$\Gamma_{ij,k} = \Gamma^l_{ij} g_{kl}. \tag{3.54}$$

Using the relation between the base vectors $\boldsymbol{g}_i$ and the Cartesian unit vectors $\boldsymbol{e}_i$,

$$\boldsymbol{g}_i = \boldsymbol{e}_m \partial_i x^m, \tag{3.55}$$

we have

$$\partial_k \boldsymbol{g}_i = \boldsymbol{e}_m \partial_k \partial_i x^m = \boldsymbol{g}_j \partial_m \xi^j \partial_k \partial_i x^m. \tag{3.56}$$

Comparing this with Eq. (3.50), we get

$$\Gamma^j_{ki} = \partial_k \partial_i x^m \partial_m \xi^j = \Gamma^j_{ik}. \tag{3.57}$$

Using $g_{ij} = \partial_i x^m \partial_j x^m$ and $\partial_i x^m \partial_m \xi^j = \delta^j_i$, we have

$$\Gamma_{ij,k} = \partial_i \partial_j x^m \partial_k x^m, \tag{3.58}$$

where the symmetry with respect to i and j is obvious. Differentiating g_{ij} and permuting the indices, we also have

$$\Gamma_{ij,k} = \frac{1}{2}(\partial_i g_{jk} + \partial_j g_{ik} - \partial_k g_{ij}). \tag{3.59}$$

Another useful result involves the derivative of $\sqrt{g}$ in the form

$$\partial_i \sqrt{g} = \partial_i L = \frac{\partial L}{\partial x^j{}_{,k}} \partial_i \partial_k x^j = L \partial_j \xi^k \partial_i \partial_k x^j = L \Gamma^k_{ik} \tag{3.60}$$

or

$$\partial_i \log \sqrt{g} = \frac{1}{\sqrt{g}} \partial_i \sqrt{g} = \Gamma^k_{ik}. \tag{3.61}$$

The parametric form of a curve in space is given by

$$\xi^i = \xi^i(a), \tag{3.62}$$

where a is a parameter. The derivative with respect to a of a vector function $\boldsymbol{v}$, defined in terms of the coordinates ξ^i, can be written as

$$\frac{\partial \boldsymbol{v}}{\partial a} = \boldsymbol{g}_j D_i v^j \frac{\partial \xi^i}{\partial a}, \tag{3.63}$$

which is known as the **absolute derivative** of the vector function.

For higher-rank tensors, we have

$$\begin{aligned} D_i A^{jk} &= \partial_i A^{jk} + \Gamma^j_{im} A^{mk} + \Gamma^k_{im} A^{jm}, \\ D_i A_{jk} &= \partial_i A_{jk} - \Gamma^m_{ij} A_{mk} - \Gamma^m_{ik} A_{jm}, \\ D_i A^j{}_k &= \partial_i A^j{}_k - \Gamma^m_{ik} A^j{}_m + \Gamma^j_{im} A^m{}_k. \end{aligned} \tag{3.64}$$

It is common practice to use the following alternative notation:

$$\begin{aligned} (\ )_{,i} &= \partial_i(\ ), \\ (\ )_{;i} &= (\ )|_i = D_i(\ ), \\ \left\{ {i \atop jk} \right\} &= \Gamma^i_{jk}, \\ \{ij, k\} &= \Gamma_{ij,k}. \end{aligned} \tag{3.65}$$

3.4 Curvature Tensors

In calculus, for a smooth function $\phi(\xi^1, \xi^2, \xi^3)$, we have

$$\partial_1 \partial_2 \phi = \partial_2 \partial_1 \phi. \tag{3.66}$$

This commutative property of partial differentiation does not apply to the covariant differential operator D_i. We could show that for any vector-valued function A_i,

$$(D_j D_i - D_i D_j) A_k = R^m_{kij} A_m, \tag{3.67}$$

where

$$R^m_{kij} = \Gamma^l_{jk} \Gamma^m_{il} - \Gamma^l_{ik} \Gamma^m_{jl} + \partial_i \Gamma^m_{jk} - \partial_j \Gamma^m_{ik}. \tag{3.68}$$

This fourth-rank tensor is known as the Riemann–Christoffel curvature tensor. If this curvature is identically zero, we can interchange the order of covariant differentiation. The curvature tensor is obtained from the derivatives of the metric tensor. If $\boldsymbol{R}$ is zero, we have a flat space. By lowering the index m, we obtain the Riemann curvature as

$$R_{ijkl} = g_{im} R^m_{jkl}. \tag{3.69}$$

In two dimensions there is only one nonvanishing component of R_{ijkl},

$$R_{1212}, \tag{3.70}$$

and in three dimensions there are six, namely,

$$R_{1212}, \quad R_{1313}, \quad R_{2323}, \quad R_{1213}, \quad R_{2123}, \quad R_{3132}. \tag{3.71}$$

These curvatures are important when we consider the *compatibility* of strain components in Chapter 5.

The Riemann curvature satisfies the **Bianchi identity**:

$$D_m R_{ijkl} + D_k R_{ijlm} + D_l R_{ijmk} = 0. \tag{3.72}$$

Two other curvature tensors of importance are the **Ricci curvature**,

$$R_{ij} = R^m_{ijm}, \tag{3.73}$$

and the **Einstein curvature**,

$$G^i_j = R^i_j - \frac{1}{2}\delta^i_j R^k{}_k, \tag{3.74}$$

which plays an important role in general relativity.

3.5 Applications

Before we leave this chapter, it is instructive to consider some examples of tensor formulations of familiar mechanics concepts.

3.5.1 Example: Incompressible Flow

In an incompressible flow the density of the fluid is constant. If we consider an infinitesimal volume in space with sides formed by the incremental curvilinear coordinates and base vectors $\boldsymbol{g}_1 d\xi^1$, $\boldsymbol{g}_2 d\xi^2$, and $\boldsymbol{g}_3 d\xi^3$ (see Fig. 3.3), the quantity of fluid entering it must be equal to the quantity leaving it.

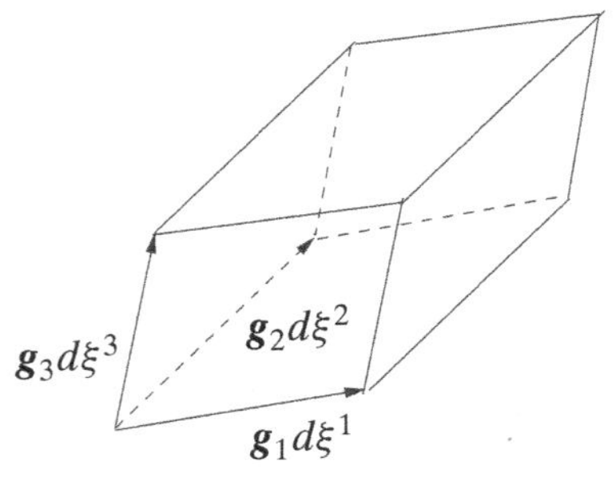

Figure 3.3. Volume element in a curvilinear system.

Let

$$\boldsymbol{v} = v^i \boldsymbol{g}_i = v_i \boldsymbol{g}^i \tag{3.75}$$

be the fluid velocity vector in the contravariant and covariant representations. Consider the vector area

$$d\boldsymbol{S}_3 = \boldsymbol{g}_1 d\xi^1 \times \boldsymbol{g}_2 d\xi^2. \tag{3.76}$$

The amount of fluid leaving this area is

$$-Q_3 = -\boldsymbol{v} \cdot d\boldsymbol{S}_3 = -v^3 \boldsymbol{g}_3 \cdot \boldsymbol{g}_1 d\xi^1 \times \boldsymbol{g}_2 d\xi^2 = -\sqrt{g} v^3 d\xi^1 d\xi^2. \tag{3.77}$$

The amount of fluid leaving the opposite surface is

$$Q_3 + dQ_3 = Q_3 + \partial_3 Q_3 d\xi^3 = Q_3 + \partial_3(\sqrt{g} v^3) d\xi^1 d\xi^2 d\xi^3. \tag{3.78}$$

For all six surfaces, the total outflow must be zero. After removing the common factor $d\xi^1 d\xi^2 d\xi^3$, we get

$$\partial_i(\sqrt{g} v^i) = 0. \tag{3.79}$$

With the gradient operator

$$\nabla = \boldsymbol{g}^i \partial_i, \tag{3.80}$$

using Eq. (3.61), we can write the **continuity** equation as

$$\nabla \cdot \boldsymbol{v} = 0. \tag{3.81}$$

3.5.2 Example: Equilibrium of Stresses

Given an infinitesimal area dS, with unit normal $\boldsymbol{n}$, the force per unit area (traction) acting on this surface is denoted by $\boldsymbol{T}^{(n)}$. As we have seen in the preceding example, we consider three special infinitesimal areas:

$$\begin{aligned} d\boldsymbol{S}_1 &= \boldsymbol{g}_2 d\xi^2 \times \boldsymbol{g}_3 d\xi^3, & dS_1 &= \sqrt{g g^{11}} d\xi^2 d\xi^3, \\ d\boldsymbol{S}_2 &= \boldsymbol{g}_3 d\xi^3 \times \boldsymbol{g}_1 d\xi^1, & dS_2 &= \sqrt{g g^{22}} d\xi^3 d\xi^1, \\ d\boldsymbol{S}_3 &= \boldsymbol{g}_1 d\xi^1 \times \boldsymbol{g}_2 d\xi^2, & dS_3 &= \sqrt{g g^{33}} d\xi^1 d\xi^2, \end{aligned} \tag{3.82}$$

$$dV = \sqrt{g} d\xi^1 d\xi^2 d\xi^3. \tag{3.83}$$

Usually there are body forces such as gravity acting on this volume. Let $\boldsymbol{f}$ be the body force per unit mass. To convert this to force per unit volume, we multiply it by the density ρ.

Balancing the forces on the six surfaces gives (after canceling dV),

$$\partial_1\left(\sqrt{g g^{11}} \boldsymbol{T}^{(1)}\right) + \partial_2\left(\sqrt{g g^{22}} \boldsymbol{T}^{(2)}\right) + \partial_3\left(\sqrt{g g^{33}} \boldsymbol{T}^{(3)}\right) + \sqrt{g} \rho \boldsymbol{f} = \boldsymbol{0}. \tag{3.84}$$

We introduce the stress tensor $\boldsymbol{\sigma}$ through the relations

$$\begin{aligned}\sqrt{g^{11}}\boldsymbol{T}^{(1)} &= \sigma^{1j}\boldsymbol{g}_j,\\ \sqrt{g^{22}}\boldsymbol{T}^{(2)} &= \sigma^{2j}\boldsymbol{g}_j,\\ \sqrt{g^{33}}\boldsymbol{T}^{(3)} &= \sigma^{3j}\boldsymbol{g}_j.\end{aligned} \tag{3.85}$$

The equilibrium of forces can now be written as

$$\partial_i(\sqrt{g}\sigma^{ij}\boldsymbol{g}_j) + \sqrt{g}\rho\boldsymbol{f} = \boldsymbol{0}. \tag{3.86}$$

Using Eq. (3.61), we may write this equation in the form

$$\nabla\cdot\boldsymbol{\sigma} + \rho\boldsymbol{f} = \boldsymbol{0} \quad \text{or} \quad D_i\sigma^{ij} + \rho f^j = 0. \tag{3.87}$$

It remains to be shown that the stress tensor transforms as it should.

SUGGESTED READING

Fung, Y. C. (1965). *Foundations of Solid Mechanics*, Prentice-Hall.
Sokolnikoff, I. S. (1951). *Tensor Analysis, Theory and Applications*, Wiley.
Synge, J. L. and Schild, A. (1949). *Tensor Calculus*, University of Toronto Press.

EXERCISES

3.1. If a curvilinear system is orthogonal, show that the metric tensor in its matrix form is diagonal.

3.2. For a curvilinear system, the gradient operator is defined as

$$\nabla = \boldsymbol{g}^i\partial_i.$$

Using this, deduce that

(a)

$$\nabla\cdot\boldsymbol{A} = D_kA^k = \frac{1}{\sqrt{g}}\frac{\partial}{\partial\xi^k}(\sqrt{g}A^k),$$

(b)

$$\begin{aligned}\nabla\cdot\nabla\phi = \nabla^2\phi &= g^{ij}D_i\partial_j\phi,\\ &= \frac{1}{\sqrt{g}}\frac{\partial}{\partial\xi^i}\left(\sqrt{g}g^{ij}\frac{\partial\phi}{\partial\xi^j}\right).\end{aligned}$$

3.3. Show that

(a)

$$D_m g_{kl} = 0,$$

(b)

$$\Gamma_{ij,k} = \frac{1}{2}(\partial_i g_{jk} + \partial_j g_{ik} - \partial_k g_{ij}),$$

(c)

$$\partial_i\boldsymbol{g}^j = -\Gamma^j_{ik}\boldsymbol{g}^k.$$

3.4. By computing

$$[D_j D_i - D_i D_j] A_k,$$

obtain an expression for the Riemann–Christoffel curvature R^m_{kij}.

3.5. A surface of revolution is given in terms of the Gaussian coordinates r and θ as

$$x = r\cos\theta,$$

$$y = r\sin\theta,$$

$$z = f(r).$$

Obtain the base vectors $\boldsymbol{g}_\alpha$ $(\alpha = 1, 2)$ and the metric tensor $g_{\alpha\beta}$. Invert the metric tensor to get $g^{\alpha\beta}$. An infinitesimal distance on this surface is given by the **first fundamental form**,

$$(ds)^2 = d\boldsymbol{r} \cdot d\boldsymbol{r} = g_{\alpha\beta} d\xi^\alpha d\xi^\beta, \quad \xi^1 = r, \quad \xi^2 = \theta.$$

The **second fundamental form** is given by

$$d\boldsymbol{r} \cdot d\boldsymbol{n} = -b_{\alpha\beta} d\xi^\alpha d\xi^\beta,$$

where $\boldsymbol{n}$ is the unit normal to the surface. Obtain $\boldsymbol{b}$ in terms of r, θ, and the derivative of f.

Find the mean and Gaussian curvatures given, respectively, by

$$\bar{\mathcal{K}} = \frac{1}{2} b^\alpha_\alpha,$$

$$\mathcal{K} = \frac{1}{g} \det \boldsymbol{b}.$$

3.6. Apply the preceding results for a sphere of radius a and for the paraboloid $f(r) = r^2/2a$, where a is a constant.

3.7. Any point on the surface of a right circular cone can be described by its distance from the apex s and the meridian angle θ. If the semiapex angle of the cone is ϕ, obtain the base vectors and the metric tensor for this coordinate system.

3.8. For a space curve with position vector $\boldsymbol{R}(s)$, where s is the curve length, the unit tangent vector $\boldsymbol{t}$, the curvature κ, the unit normal $\boldsymbol{n}$, the binormal $\boldsymbol{b}$, and the torsion τ are given by

$$\boldsymbol{t} = \frac{d\boldsymbol{R}}{ds}, \quad \kappa\boldsymbol{n} = \frac{d\boldsymbol{t}}{ds}, \quad \boldsymbol{b} = \boldsymbol{t} \times \boldsymbol{n}, \quad \tau\boldsymbol{n} = -\frac{d\boldsymbol{b}}{ds}.$$

Show that these quantities satisfy the Frenet–Serret formula,

$$\frac{d\boldsymbol{n}}{ds} = -\kappa\boldsymbol{t} + \tau\boldsymbol{b}.$$

Illustrate this for the helix given by

$$x^1 = a\cos\theta, \quad x^2 = a\sin\theta, \quad x^3 = b\theta,$$

where a and b are constants.

3.9. In polar coordinates, the unit vectors $\boldsymbol{e}_r$ and $\boldsymbol{e}_\theta$ are functions of θ and the gradient operator is given by

$$\nabla = \boldsymbol{e}_r \frac{\partial}{\partial r} + \boldsymbol{e}_\theta \frac{\partial}{r\partial\theta}.$$

In the 2D case the physical components of the stress are

$$\boldsymbol{\sigma} = \begin{bmatrix} \sigma_{rr} & \sigma_{r\theta} \\ \sigma_{r\theta} & \sigma_{\theta\theta} \end{bmatrix}.$$

Expand the equilibrium equation

$$\nabla \cdot \boldsymbol{\sigma} + \rho \boldsymbol{f} = \mathbf{0}.$$

3.10. For the polar coordinates,

$$r = \sqrt{x_1^2 + x_2^2}, \quad \theta = \arctan(x_2/x_1).$$

Obtain the base vectors, metric tensor, and the gradient operator for this system.

3.11. The spherical coordinates r, θ, and ϕ are related to the Cartesian coordinates, as

$$x^1 = r \sin\theta \cos\phi,$$

$$x^2 = r \sin\theta \sin\phi,$$

$$x^3 = r \cos\theta.$$

Compute the base vectors, metric tensor, and gradient operator for the spherical coordinate system.

3.12. The base and a side of a parallelogram make an acute angle α. If a 2D system of coordinates is constructed in such a way that ξ^1 is parallel to the base and ξ^2 is parallel to the side, find a relation between this system and the Cartesian system. Obtain the base vectors, metric tensor, and Laplace operator for this system. (In the analysis of plates that are parts of a swept-wing airplane, parallelogram shapes are encountered.)

3.13. The Cartesian coordinates (x, y) and the "pseudoelliptical" coordinates (ξ, η) are related by

$$x = a\xi \cos\eta, \quad y = b\xi \sin\eta,$$

where a and b are constants.
Compute

(a) the base vectors,
(b) the metric tensor,
(c) the reciprocal base vectors,
(d) the reciprocal metric tensor, and
(e) the Laplace operator.

3.14. In the context of the examples of this chapter, referring to Fig. 3.4, let

$$\overrightarrow{OA} = \boldsymbol{r}_1 = \boldsymbol{g}_1 d\xi^1, \quad \overrightarrow{OB} = \boldsymbol{r}_2 = \boldsymbol{g}_2 d\xi^2, \quad \overrightarrow{OC} = \boldsymbol{r}_3 = \boldsymbol{g}_3 d\xi^3.$$

Using the cross products of vectors, show that the vector areas satisfy

$$\overrightarrow{ABC} = \overrightarrow{OAB} + \overrightarrow{OBC} + \overrightarrow{OCA},$$

where $\overrightarrow{ABC}$ has its normal outward from the volume element and the areas on the right-hand side have their normals inward. If these areas have magnitudes dS, dS_3, dS_1, and dS_2, respectively, and unit normals $\boldsymbol{n}$, $\boldsymbol{n}_3$, $\boldsymbol{n}_1$, and $\boldsymbol{n}_2$, show that

$$\boldsymbol{n} dS = \boldsymbol{n}_i dS_i.$$

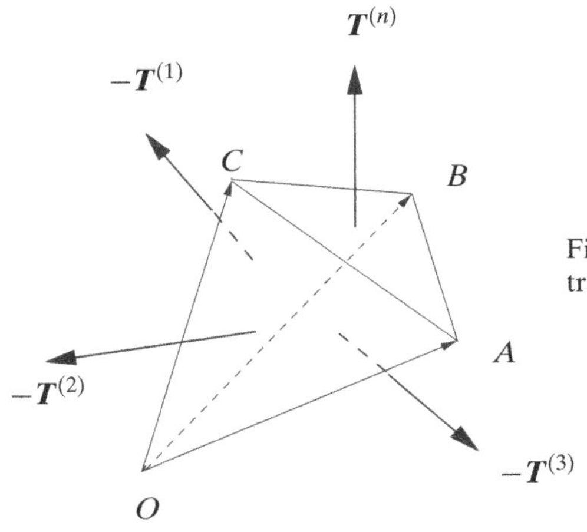

Figure 3.4. Tetragonal volume element with traction vectors.

If the traction acting on dS is $\boldsymbol{T}^{(n)}$, etc., in the limit as the volume of the tetrahedron goes to zero, show that

$$\boldsymbol{T}^{(n)} dS = \boldsymbol{T}^{(i)} dS_i.$$

3.15. In the preceding exercise, if the unit vector $\boldsymbol{n}$ is resolved as $\boldsymbol{n} = n_i \boldsymbol{g}^i$ and introducing the stress tensor σ^{ij} as $\boldsymbol{T}^{(i)}\sqrt{g^{ii}} = \sigma^{ij}\boldsymbol{g}_j$ (no sum on i), show that

$$T^{(n)j} = \sigma^{ij} n_i.$$

3.16. Obtain the characteristic equation for the principal stresses by extremizing $\boldsymbol{T}^{(n)} \cdot \boldsymbol{n}$.

3.17. In Exercise 3.14, if we introduce a new coordinate system $\bar{\xi}^i$ in such a way that the same area element ABC caps the three vectors,

$$\overrightarrow{\bar{O}A} = \bar{\boldsymbol{g}}_1 d\bar{\xi}^1, \quad \text{etc.,}$$

where $\bar{O}$ is distinct from O, show that

$$\boldsymbol{T}^{(i)} dS_i = \bar{\boldsymbol{T}}^{(i)} d\bar{S}_i.$$

Using the preceding results, prove that the stress tensor σ^{ij} obeys the transformation law for a second-rank contravariant tensor.

3.18. In dynamics the Lagrangian L is defined as

$$L = K - V,$$

where K is the kinetic energy and V is the potential of the forces. For a single particle of mass m moving in a curvilinear coordinate system without any external forces, obtain the equation of motion from

$$\frac{d}{dt}\frac{\partial L}{\partial \dot{x}^i} - \frac{\partial L}{\partial x^i} = 0,$$

where $x^i(t)$ are the curvilinear coordinates of the particle.

4 Integral Theorems

To deal with the conservation and balance laws of mechanics, we use the integral theorems of Gauss and Stokes frequently. Although most undergraduate calculus courses cover this material, it is of interest to reconsider these theorems by using the index notation and the summation convention.

4.1 Gauss Theorem

Consider a convex region V bounded by a smooth surface S in three dimensions. For a nonconvex region, if it can be divided into a finite number of convex subregions, we could still use the following theorem. Let $A(x_1, x_2, x_3)$ be a differentiable (as many times as we need) function defined in V. We begin by considering the integral

$$I = \iiint_V \frac{\partial A}{\partial x_1} dx_1 dx_2 dx_3. \tag{4.1}$$

As shown in Fig. 4.1, a differential tube of cross-sectional area $dx_2 dx_3$ intersects the surface S at two places, dividing the total surface into S^* and S^{**}. Integrating with respect to x_1, we have

$$I = \iint_{S^*} A dx_2 dx_3 - \iint_{S^{**}} A dx_2 dx_3. \tag{4.2}$$

With

$$dx_2 dx_3 = n_1^* dS \quad \text{on} \quad S^*, \quad dx_2 dx_3 = -n_1^{**} dS \quad \text{on} \quad S^{**}, \tag{4.3}$$

we get

$$I = \int_S A n_1 dS. \tag{4.4}$$

Next, we generalize this to get

$$\int_V \partial_i A dV = \int_S n_i A dS. \tag{4.5}$$

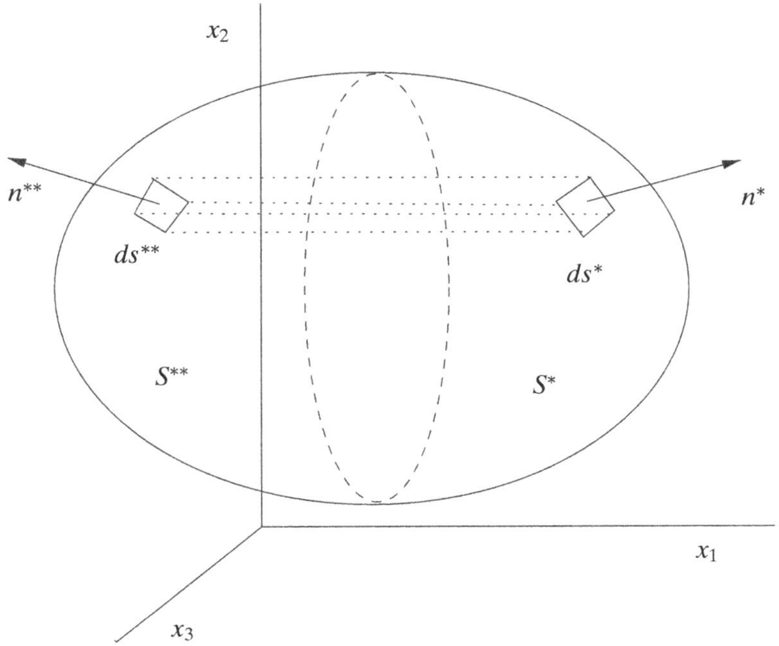

Figure 4.1. Integration in three dimensions.

If there are a finite number of internal surfaces in V where A is not differentiable, but left and right limits for A exist as these surfaces are crossed, the contributions from these surfaces have to be added to the preceding result. Such discontinuous behavior occurs in situations involving shock waves and fracture surfaces in solids.

If we replace A with a component of a tensor, $A_{jkl}\ldots$, we get

$$\int_V \partial_i A_{jkl}\ldots dV = \int_S n_i A_{jkl\ldots} dS. \tag{4.6}$$

As special cases, we have the following relations:

1.

$$\int_V \partial_i A_i dV = \int_S n_i A_i dS \quad \text{or} \quad \int_V \nabla \cdot \mathbf{A} dV = \int_S \boldsymbol{n} \cdot \mathbf{A} dS, \tag{4.7}$$

which is the well-known divergence theorem.

2.

$$\int_V \partial_i A dV = \int_S n_i A dS \quad \text{or} \quad \int_V \nabla A dV = \int_S \boldsymbol{n} A dS, \tag{4.8}$$

where the equation on the left has been multiplied by $\boldsymbol{e}_i$ to get the vector form of the equation on the right.

3.

$$\int_V \partial_i A_j dV = \int_S n_i A_j dS \quad \text{or} \quad \int_V \nabla \times \mathbf{A} dV = \int_S \boldsymbol{n} \times \mathbf{A} dS, \tag{4.9}$$

where we have used multiplication by $e_{ijk}\mathbf{e}_k$.

The preceding results can be summarized as

$$\int_V \nabla * \boldsymbol{A} dV = \int_S \boldsymbol{n} * \boldsymbol{A} dS, \tag{4.10}$$

where $*$ stands for $\cdot$, $\times$, or $\otimes$, producing the dot product, cross product, or the tensor product, and $\boldsymbol{A}$ is a tensor of any rank. The Gauss theorem has also been attributed to Green and Ostrogradsky.

We can also see that the preceding result applies to converting area integrals into contour integrals, instead of volume integrals into surface integrals.

4.2 Stokes Theorem

Consider a two-dimensional (2D) convex region S bounded by the curve C in the x_1, x_2 plane (see Fig. 4.2). We assume A is differentiable inside S. The area integral

$$I = \iint_S \frac{\partial A}{\partial x_1} dx_1 dx_2 \tag{4.11}$$

can be evaluated by integrating with respect to x_1 as

$$I = \int_{C^*} A dx_2 - \int_{C^{**}} A dx_2 = \oint_C A dx_2. \tag{4.12}$$

Here, we have shown the two points P_1 and P_2, which are located at the minimum and maximum values of x_2, dividing the curve C into C^* and C^{**}. Similarly, for a function B,

$$J = \iint_S \frac{\partial B}{\partial x_2} dx_1 dx_2 = \int_{C'} B dx_1 - \int_{C''} B dx_1 = -\oint_C B dx_1, \tag{4.13}$$

where we find C'' and C' by dividing the curve C, using the minimum and maximum values of x_1. Combining I and J, we obtain

$$\iint_S \left[\frac{\partial A}{\partial x_1} - \frac{\partial B}{\partial x_2} \right] dx_1 dx_2 = \oint_C [A dx_2 + B dx_1]. \tag{4.14}$$

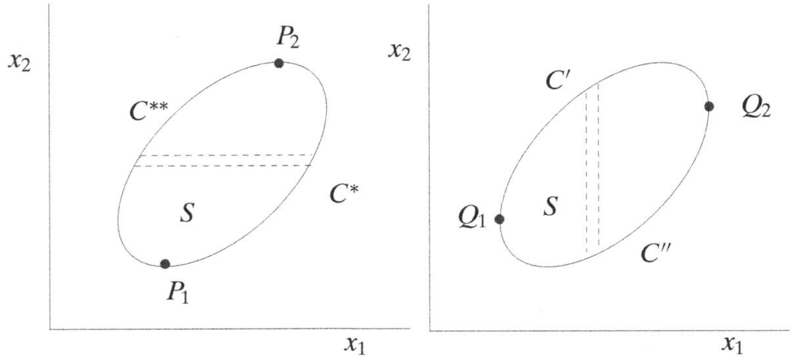

Figure 4.2. Integration in two dimensions.

If we equate A and B to vector components,

$$A = A_2, \quad B = A_1, \tag{4.15}$$

we find

$$\iint_S \left[\frac{\partial A_2}{\partial x_1} - \frac{\partial A_1}{\partial x_2}\right] dx_1 dx_2 = \oint_C [A_1 dx_1 + A_2 dx_2]. \tag{4.16}$$

This result is known as the Stokes theorem. The left-hand side of this equation has the $\boldsymbol{e}_3$ component of the *curl* of the vector **A**. We may write this as

$$\iint_S \boldsymbol{e}_3 \cdot \nabla \times \mathbf{A} dx_1 dx_2 = \oint_C \mathbf{A} \cdot d\boldsymbol{x}. \tag{4.17}$$

This shows that the integral of the normal component of the *curl* of a vector field on the surface S is equal to the integral of the tangential component of the same field around the closed curve C.

If we have a curved surface S in three dimensions bounded by a curve C, can we relate the normal component of the curl to the tangential component as we have done previously? The answer is "yes," provided there is a mapping that transforms the curved surface in three dimensions to a flat surface in two dimensions.

Let us assume a 2D plane (y_1, y_2), with every point in a convex region $\bar{S}$ mapped onto a point (x_1, x_2, x_3), with mapping functions

$$x_i = x_i(y_1, y_2). \tag{4.18}$$

Then $\bar{S}$ would be mapped into a surface S, as shown in Fig. 4.3. The boundary curve $\bar{C}$ would be mapped into C. Assume that there is a vector-valued function

$$\bar{\mathbf{A}} = \bar{\mathbf{A}}(y_1, y_2) \tag{4.19}$$

that satisfies the Stokes theorem,

$$\iint_{\bar{S}} \left[\frac{\partial \bar{A}_2}{\partial y_1} - \frac{\partial \bar{A}_1}{\partial y_2}\right] dy_1 dy_2 = \oint_{\bar{C}} [\bar{A}_1 dy_1 + \bar{A}_2 dy_2]. \tag{4.20}$$

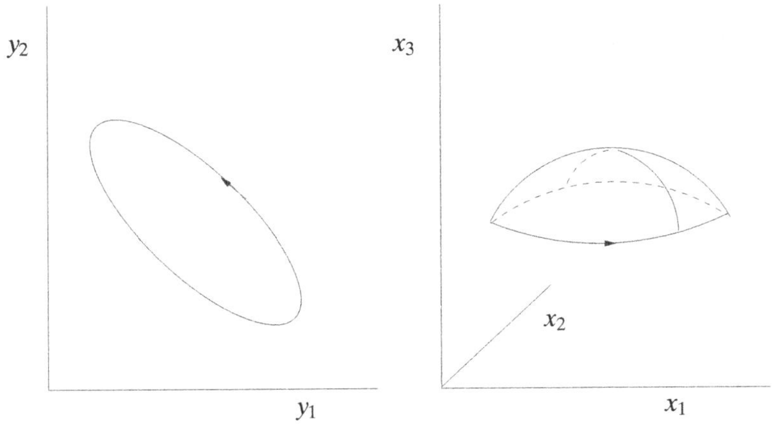

Figure 4.3. Mapping a 2D surface onto a 3D surface.

To distinguish the partial derivatives with respect to the y coordinates from those with respect to the x coordinates, we use the notation

$$\partial_\alpha = \frac{\partial}{\partial y_\alpha}, \quad \alpha = 1, 2; \quad \partial_i = \frac{\partial}{\partial x_i}, \quad i = 1, 2, 3. \tag{4.21}$$

Next, consider a three-vector $A_i(x_1, x_2, x_3)$, defined on the surface S, which has unit normal $\boldsymbol{n}$. Let us compute

$$I = \int_S \boldsymbol{n} \cdot \nabla \times \mathbf{A} dS = \int_S n_i \mathbf{e}_i \cdot e_{jkl} \partial_k A_l \mathbf{e}_j dS = \int_S e_{ikl} n_i \partial_k A_l dS. \tag{4.22}$$

Let us add a y_3 coordinate and three unit vectors $\bar{\mathbf{e}}_1$, $\bar{\mathbf{e}}_3$, and $\bar{\mathbf{e}}_3$ to our y_1, y_2 plane.

We know that the surface area dS with unit normal $\boldsymbol{n}$ is the map of the area $dy_1 dy_2$ with unit normal $\bar{\boldsymbol{e}}_3$. Let $d\boldsymbol{x}$ be the map of $dy_1 \bar{\boldsymbol{e}}_1$ and $d\boldsymbol{x}^*$ be the map of $dy_2 \bar{\boldsymbol{e}}_2$. We have

$$d\bar{S}\boldsymbol{e}_3 = dy_1 \bar{\boldsymbol{e}}_1 \times dy_2 \bar{\boldsymbol{e}}_2, \quad d\bar{S} = dy_1 dy_2. \tag{4.23}$$

The map of this **vector area** can be written as

$$\boldsymbol{n} dS = d\boldsymbol{x} \times d\boldsymbol{x}^* \tag{4.24}$$

or

$$\boldsymbol{n} dS = \partial_1 x_i \boldsymbol{e}_i dy_1 \times \partial_2 x_j \boldsymbol{e}_j dy_2. \tag{4.25}$$

After the cross product is performed,

$$\boldsymbol{n} dS = \partial_1 x_i \partial_2 x_j e_{ijk} \boldsymbol{e}_k dy_1 dy_2. \tag{4.26}$$

The i component of this vector is

$$n_i dS = \partial_1 x_j \partial_2 x_k e_{ijk} dy_1 dy_2. \tag{4.27}$$

We substitute this result into the expression for I to get

$$I = \int_{\bar{S}} e_{ikl} \partial_k A_l \partial_1 x_j \partial_2 x_m e_{ijm} dy_1 dy_2. \tag{4.28}$$

Using the e–δ identity, we can simplify this to get

$$I = \int_{\bar{S}} [\partial_k A_l - \partial_l A_k] \partial_1 x_k \partial_2 x_l dy_1 dy_2. \tag{4.29}$$

So far we have kept two sets of functions, $\bar{A}_\alpha$ and A_i. If we relate these two sets as

$$\bar{A}_\alpha = (\partial_\alpha x_i) A_i, \tag{4.30}$$

we have

$$\partial_1 \bar{A}_2 = (\partial_1 \partial_2 x_i) A_i + \partial_j A_i \partial_2 x_i \partial_1 x_j, \tag{4.31}$$

$$\partial_2 \bar{A}_1 = (\partial_2 \partial_1 x_i) A_i + \partial_j A_i \partial_1 x_i \partial_2 x_j, \tag{4.32}$$

$$I = \int_{\bar{S}} [\partial_1 \bar{A}_2 - \partial_2 \bar{A}_1] dy_1 dy_2. \tag{4.33}$$

Finally,

$$\int_S \boldsymbol{n} \cdot \nabla \times \boldsymbol{A} dS = \oint_C d\boldsymbol{x} \cdot \boldsymbol{A}. \tag{4.34}$$

This result is known as the Kelvin transformation. Note that, in writing $\boldsymbol{A}$ on the right of the differential in the line integral, the Stokes theorem with the Kelvin transformation has been made applicable to tensors of ranks higher than one. The concept of mapping previously introduced anticipates deformation of bodies presented in later chapters.

SUGGESTED READING

Greenberg, M. D. (1988). *Advanced Engineering Mathematics*, Prentice-Hall.
Kreyzig, E. (2006). *Advanced Engineering Mathematics*, 9th ed. Wiley.
Sokolnikoff, I. S. and Redheffer, R. M. (1966). *Mathematics of Physics and Modern Engineering*, McGraw-Hill.
Thomas, G. B. and Finney, R. L. (1979). *Calculus and Analytic Geometry*, Addison-Wesley.

EXERCISES

4.1. Prove the Green's theorem:

$$\int_V [u\nabla^2 v - v\nabla^2 u] dV = \int_S [u\nabla v - v\nabla u] \cdot \boldsymbol{n} dS.$$

4.2. If $\boldsymbol{a}$ is a constant vector, show that

$$\oint_C \boldsymbol{a} \cdot d\boldsymbol{x} = 0, \quad \oint_C \boldsymbol{a} \times d\boldsymbol{x} = \boldsymbol{0}.$$

4.3. Transform the surface integrals

$$I = \int_S (\boldsymbol{n} \times \nabla) \cdot \boldsymbol{u} dS, \quad \boldsymbol{I} = \int_S (\boldsymbol{n} \times \nabla) \times \boldsymbol{u} dS$$

into volume integrals and evaluate them.

4.4. Obtain the differential equations for the vector function $\boldsymbol{\phi}$ and the scalar ψ inside an arbitrary volume if the surface integrals

$$\int_S \boldsymbol{n} \times (r\boldsymbol{\phi}) da = \boldsymbol{0}, \quad \int_S \boldsymbol{n} \cdot (\boldsymbol{r}\psi) da = 0,$$

where $r = (x_i x_i)^{1/2}$ and $\boldsymbol{r} = \boldsymbol{x}$.

4.5. In the torsion of shafts, the St. Venant warping function ϕ satisfies the Laplace equation in a simply connected 2D domain D in the x, y plane, representing the cross-section of the shaft and the boundary conditions,

$$\boldsymbol{n} \cdot \nabla\phi = \boldsymbol{k} \cdot \boldsymbol{n} \times \boldsymbol{r},$$

where $\boldsymbol{n}$ is the unit normal to the boundary, $\boldsymbol{r} = x\boldsymbol{i} + y\boldsymbol{j}$, and $\boldsymbol{k}$ is the unit vector perpendicular to the plane.
Show the integrals representing the shear forces in the x and y directions,

$$\int_D \left(\frac{\partial\phi}{\partial x} - y\right) dA = 0, \quad \int_D \left(\frac{\partial\phi}{\partial y} + x\right) dA = 0.$$

4.6. If

$$v = e_3 \times r/r^2, \quad r = x_1 e_1 + x_2 e_2, \quad r = |r|,$$

compute

$$I = \oint v \cdot dx$$

around a unit circle, $r = 1$, directly. Compare your result with the value obtained with the Stokes theorem. Explain the reason for the difference.

4.7. In an infinite 3D medium, the function u satisfies Poisson's equation,

$$\nabla^2 u = q(x),$$

with $u \to 0$ as $r = |x| \to \infty$. To obtain a general solution, we use the function

$$v(x) = \frac{1}{4\pi |x - x'|}.$$

Show that v satisfies the Laplace equation everywhere except at $x = x'$. Remove a sphere of radius ϵ centered at $x = x'$ and apply the Green's theorem to u and v with the volume integral over the infinite domain without the small sphere. In the limit $\epsilon \to 0$, show that

$$u(x) = -\frac{1}{4\pi} \int_V \frac{q(x')dV}{|x' - x|},$$

where V is the infinite domain. Here, v is called the Green's function for Poisson's equation.

4.8. Consider an arbitrary simple closed curve C inside a domain D in the complex plane $z = x + iy$. For a complex valued function, $w(z) = u(x, y) + iv(x, y)$, if

$$\oint_C w dz = 0$$

for all such curves, show that

$$\frac{\partial u}{\partial x} = \frac{\partial v}{\partial y}, \quad \frac{\partial u}{\partial y} = -\frac{\partial v}{\partial x}.$$

These are known as the Cauchy–Riemann equations. This result is known as Morera's theorem.

4.9. In the 2D case, derive the Stokes theorem using the Gauss theorem.

5 Deformation

Consider a material continuum occupying a region B_0 with volume V_0 and surface area S_0 at time $t = 0$. At a later time t, we see this body in a region B with volume V and surface area S. We refer to these two configurations as **undeformed** and **deformed configurations**, respectively. We use a fixed Cartesian system to describe the geometry of the two configurations. As shown in Fig. 5.1, a generic particle in the undeformed configuration has the position $\boldsymbol{X}$, and the same particle in the deformed configuration has the position $\boldsymbol{x}$. We simultaneously develop descriptions of the geometry of this mapping from the state B_0 to B, using X_1, X_2, and X_3 as independent variables and using x_1, x_2, and x_3 as independent variables. To distinguish the partial derivative operator ∂ in the two systems, we use Latin indices, $i, j, k, \ldots$, for the components of $\boldsymbol{x}$ and Greek indices, $\alpha, \beta, \gamma, \ldots$, for the components of $\boldsymbol{X}$. For the indices of the unit vectors to match the components of the vectors, we denote $\boldsymbol{e}_1$, $\boldsymbol{e}_2$, and $\boldsymbol{e}_3$ by $\boldsymbol{e}_i$ in one system and by $\boldsymbol{e}_\alpha$ in the other. Thus

$$\boldsymbol{X} = X_\alpha \boldsymbol{e}_\alpha, \qquad \boldsymbol{x} = x_i \boldsymbol{e}_i. \tag{5.1}$$

We introduce the partial derivatives

$$\partial_\alpha = \frac{\partial}{\partial X_\alpha}, \qquad \partial_i = \frac{\partial}{\partial x_i}. \tag{5.2}$$

5.1 Lagrangian and Eulerian Descriptions

If we use X_α as independent variables and x_i as dependent variables, i.e.,

$$x_i = x_i(X_1, X_2, X_3) = x_i(\boldsymbol{X}), \tag{5.3}$$

we have the Lagrangian description of deformation. The variables X_α are called the material coordinates or the Lagrangian coordinates.

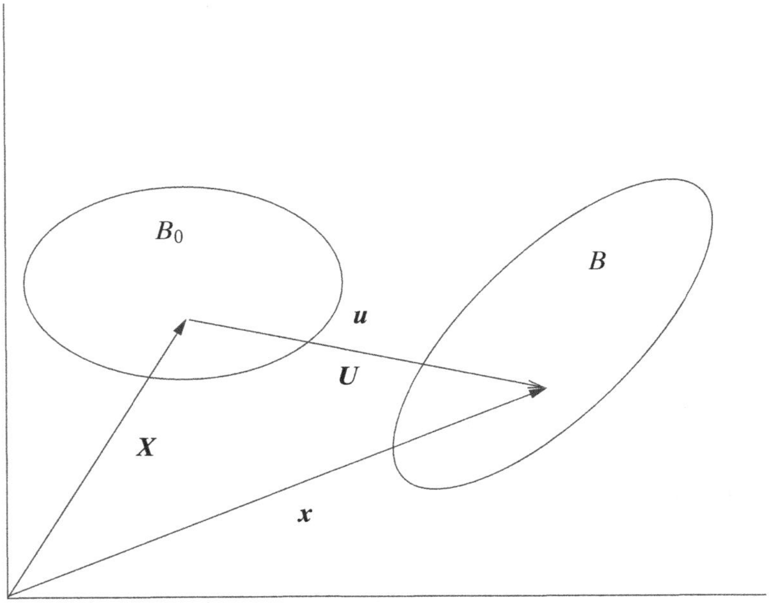

Figure 5.1. Deformation of a body.

If x_i are the independent variables and X_α are the dependent variables, i.e.,

$$X_\alpha = X_\alpha(x_1, x_2, x_3) = X_\alpha(\boldsymbol{x}), \tag{5.4}$$

we have the Eulerian description of deformation. The coordinates x_i are called the spatial coordinates or the Eulerian coordinates.

Depending on the complexity of the differential equations, the boundary conditions, and the usefulness of the solution variable, one of these formulations may be preferable over the other in a particular application. For example, in the mechanics of solids, when we load a rectangular, uniformly thick plate, it deforms into a complex curved shape. From the simplicity of the initial geometry, we may prefer the Lagrangian formulation for this case. For a fluid flow problem, we can observe the flow characteristics through a wind-tunnel window. We may not care where the various fluid particles were at time $t = 0$. An Eulerian approach is often used in fluid dynamics.

In the Lagrangian formulation, we label a material particle and follow it as it moves with time. In Eulerian formulation, we stand at a spatial point and observe the passing particles. As $\boldsymbol{X}$ and $\boldsymbol{x}$ are locations of the same material particle, the functional relations, Eqs. (5.1), can be inverted in the neighborhood of a point $\boldsymbol{X}$ to obtain Eqs. (5.2). This one-to-one correspondence is known as the **axiom of continuity**. This axiom implies (a) the indestructible nature of matter and (b) the impenetrability of matter. According to (a), matter occupying a finite volume cannot be deformed into zero volume, and, according to (b), the motion of a body carries every material line into another line and every surface into another surface. A smooth curve cannot be deformed into an intersecting curve.

The necessary and sufficient condition for the existence of the one-to-one correspondence of the functions X_α and x_i in a small neighborhood is

$$\det[\partial_\alpha x_i] \neq 0. \tag{5.5}$$

The displacement vector is defined as the vectorial distance from the initial position $\boldsymbol{X}$ to the final position $\boldsymbol{x}$ of a particle. When this vector is considered to be a function of the material coordinates we denote it by $\boldsymbol{U}$, and when it is a function of spatial coordinates we use $\boldsymbol{u}$:

$$\boldsymbol{U}(X_1, X_2, X_3) = \boldsymbol{u}(x_1, x_2, x_3) = \boldsymbol{x} - \boldsymbol{X}. \tag{5.6}$$

In component form,

$$U_\alpha = x_i \delta_{i\alpha} - X_\alpha, \qquad u_i = x_i - X_\alpha \delta_{\alpha i}, \tag{5.7}$$

where, for consistency of notation, the Kronecker delta is introduced.

5.2 Deformation Gradients

From Eqs. (5.3) and (5.4) we get

$$dx_i = dX_\alpha \partial_\alpha x_i, \quad dX_\alpha = dx_i \partial_i X_\alpha. \tag{5.8}$$

The quantities $\partial_\alpha x_i$ and their inverses $\partial_i X_\alpha$ are called the deformation gradients. Figure 5.2 shows the mapping of an element, dX_1. Differentiating Eq. (5.3) by x_j and Eq. (5.4) by X_β, we see

$$\partial_\alpha x_i \partial_j X_\alpha = \delta_{ij}, \quad \partial_i X_\alpha \partial_\beta x_i = \delta_{\alpha\beta}. \tag{5.9}$$

Let us introduce the Green's deformation gradient tensor $\boldsymbol{F}$ and the Cauchy deformation gradient tensor $\boldsymbol{f}$, in matrix forms:

$$\boldsymbol{F} = \nabla \boldsymbol{x} = [\partial_\alpha x_i], \quad \boldsymbol{f} = \nabla \boldsymbol{X} = [\partial_i X_\alpha]. \tag{5.10}$$

These matrices are inverses of each other:

$$\boldsymbol{F}\boldsymbol{f} = \boldsymbol{I} = \boldsymbol{f}\boldsymbol{F}. \tag{5.11}$$

It should be noted that in the literature the transposes of these matrices are often used for the deformation gradients.

The axiom of continuity requires that these matrices be nonsingular. The determinants of these matrices are the Jacobian determinants of transformations (5.3) and (5.4):

$$J = |\partial_\alpha x_i|, \quad j = |\partial_i X_\alpha|, \quad Jj = 1. \tag{5.12}$$

There will not be any problem in distinguishing the Jacobian "j" from the integer indices.

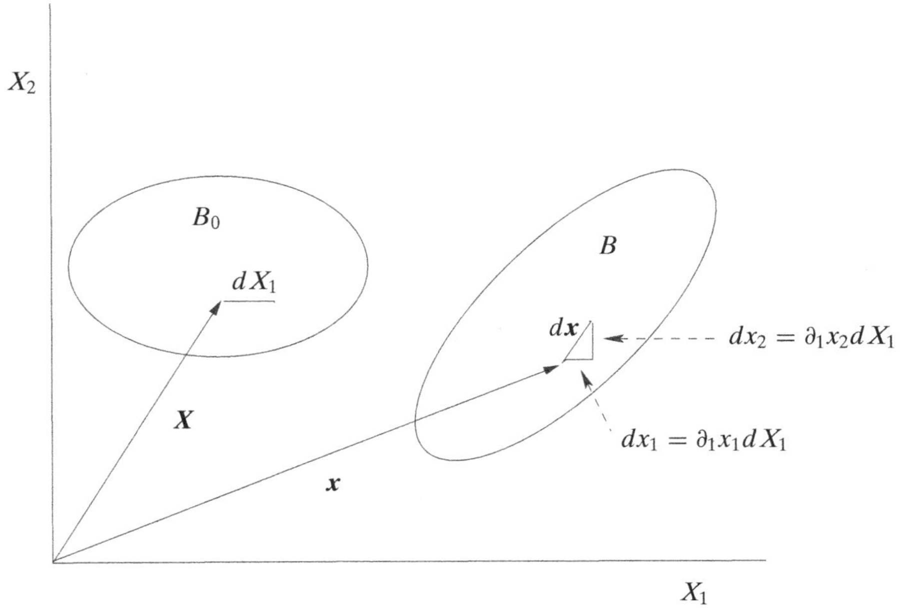

Figure 5.2. Deformation of an element.

These determinants have the expansions

$$J = \frac{1}{6} e_{ijk} e_{\alpha\beta\gamma} \partial_\alpha x_i \partial_\beta x_j \partial_\gamma x_k, \quad j = \frac{1}{6} e_{\alpha\beta\gamma} e_{ijk} \partial_i X_\alpha \partial_j X_\beta \partial_k X_\gamma. \tag{5.13}$$

Differentiating J by $F_{\alpha i} = \partial_\alpha x_i$ and j by $f_{i\alpha} = \partial_i X_\alpha$, we get

$$\frac{\partial J}{\partial F_{\alpha i}} = \frac{1}{2} e_{ijk} e_{\alpha\beta\gamma} \partial_\beta x_j \partial_\gamma x_k = \text{cofactor of } \partial_\alpha x_i, \tag{5.14}$$

$$\frac{\partial j}{\partial f_{i\alpha}} = \frac{1}{2} e_{\alpha\beta\gamma} e_{ijk} \partial_j X_\beta \partial_k X_\gamma = \text{cofactor of } \partial_i X_\alpha. \tag{5.15}$$

Remembering the relation between cofactors and inverses, we can write $\partial_i X_\alpha$ in terms of $\partial_\alpha x_i$, and vice versa, as

$$\partial_i X_\alpha = \frac{1}{2J} e_{ijk} e_{\alpha\beta\gamma} \partial_\beta x_j \partial_\gamma x_k, \tag{5.16}$$

$$\partial_\alpha x_i = \frac{1}{2j} e_{\alpha\beta\gamma} e_{ijk} \partial_j X_\beta \partial_k X_\gamma. \tag{5.17}$$

5.2.1 Deformation Gradient Vectors

We know that an element $d\boldsymbol{X}$ gets mapped into $d\boldsymbol{x}$ during deformation. In particular, the element $\boldsymbol{e}_1 dX_1$ goes to

$$d\boldsymbol{x} = \partial_1 x_i \boldsymbol{e}_i dX_1 = \boldsymbol{G}_1 dX_1, \quad \boldsymbol{G}_1 = \partial_1 x_i \boldsymbol{e}_i. \tag{5.18}$$

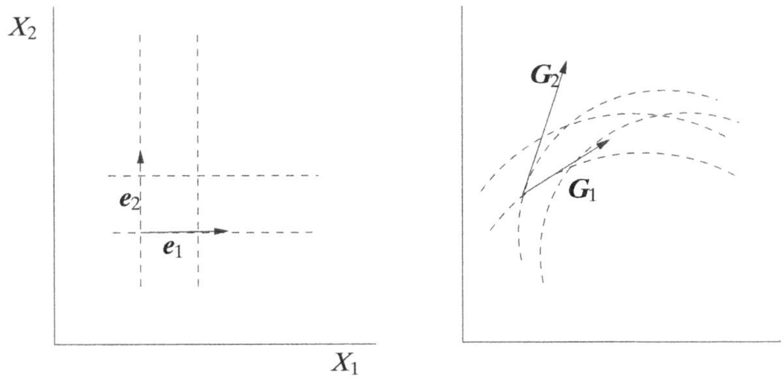

Figure 5.3. Deformation gradient vectors $\boldsymbol{G}_1$ and $\boldsymbol{G}_2$.

If we had marked $\boldsymbol{e}_1$ in the undeformed configuration, we would see its image $\boldsymbol{G}_1$ in the deformed configuration. The three unit vectors have

$$\boldsymbol{G}_\alpha = \partial_\alpha x_i \boldsymbol{e}_i = \partial_\alpha \boldsymbol{x} \tag{5.19}$$

as their images. These are known as the Green's deformation gradient vectors (see Fig. 5.3).

Similarly, the unit vectors $\boldsymbol{e}_i$ in the current configuration had their images

$$\boldsymbol{g}_i = \partial_i X_\alpha \boldsymbol{e}_\alpha = \partial_i \boldsymbol{X} \tag{5.20}$$

in the undeformed configuration. These are called the Cauchy deformation gradient vectors.

These two systems of vectors are intimately related to the base vectors we had seen in Chapter 3. To see this, consider an arbitrary element $d\boldsymbol{X}$ getting mapped into $d\boldsymbol{x}$. We have

$$d\boldsymbol{x} = \partial_\alpha \boldsymbol{x} dX_\alpha = \boldsymbol{G}_\alpha dX_\alpha, \tag{5.21}$$

$$d\boldsymbol{X} = \partial_i \boldsymbol{X} dx_i = \boldsymbol{g}_i dx_i. \tag{5.22}$$

The lengths of these elements can be expressed as

$$(dS)^2 = dX_\alpha dX_\alpha, \quad (ds)^2 = dx_i dx_i. \tag{5.23}$$

If we write the current length in terms of material coordinates, we have

$$(ds)^2 = \partial_\alpha \boldsymbol{x} dX_\alpha \cdot \partial_\beta \boldsymbol{x} dX_\beta = \boldsymbol{G}_\alpha \cdot \boldsymbol{G}_\beta dX_\alpha dX_b = G_{\alpha\beta} dX_\alpha dX_\beta, \tag{5.24}$$

where $G_{\alpha\beta}$ is our old metric tensor, which is the dot product of the base vectors. In continuum mechanics we refer to $\boldsymbol{G} = [G_{\alpha\beta}]$ as the Green's deformation tensor.

A parallel consideration in terms of spatial coordinates gives

$$(dS)^2 = g_{ij} dx_i dx_j, \quad g_{ij} = \boldsymbol{g}_i \cdot \boldsymbol{g}_j. \tag{5.25}$$

The tensor $\boldsymbol{g}$ is called the Cauchy deformation tensor. We may express the deformation tensors, $\boldsymbol{G}$ and $\boldsymbol{g}$, as

$$\boldsymbol{G} = \boldsymbol{F}\boldsymbol{F}^T, \quad \boldsymbol{g} = \boldsymbol{f}\boldsymbol{f}^T. \tag{5.26}$$

The tensors $\boldsymbol{G}$ and $\boldsymbol{g}$ are symmetric by virtue of their definition in terms of dot products. Because the squares of distances are positive, these tensors (matrices) are positive definite.

5.2.2 Curvilinear Systems

The base vectors $\boldsymbol{G}_\alpha$ or $\boldsymbol{g}_i$ remind us of curvilinear coordinate systems. If we had etched a rectilinear grid on the undeformed body, after deformation, we would see the grid deformed into a curvilinear net. This is the curvilinear system associated with base vectors $\boldsymbol{G}_\alpha$ and metric tensor $\boldsymbol{G}$. In the context of general tensors, it is appropriate to use X^α and x^i as the coordinates. We can imagine a similar picture for a rectangular grid in the current configuration.

We may introduce reciprocal base vectors, $\boldsymbol{G}^\alpha$ and $\boldsymbol{g}^i$, to go with the curvilinear coordinates. These are defined as

$$\boldsymbol{G}^\alpha = \boldsymbol{e}_i \partial_i X^\alpha, \quad \boldsymbol{g}^i = \boldsymbol{e}_\alpha \partial_\alpha x^i. \tag{5.27}$$

Using the reciprocity of $\boldsymbol{F}$ and $\boldsymbol{f}$, we see that

$$\boldsymbol{G}^\alpha \cdot \boldsymbol{G}_\beta = \delta^\alpha_\beta, \quad \boldsymbol{g}^i \cdot \boldsymbol{g}_j = \delta^i_j. \tag{5.28}$$

Associated with the reciprocal base vectors are the reciprocal deformation tensors $G^{\alpha\beta}$ and g^{ij}, which are inverses of $G_{\alpha\beta}$ and g_{ij}, respectively. In matrix form,

$$\begin{aligned}
\boldsymbol{G}^{-1} &= [G^{\alpha\beta}] = [\partial_i X^\alpha \partial_i X^\beta] = \boldsymbol{f}^T\boldsymbol{f}, \\
\boldsymbol{G} &= [G_{\alpha\beta}] = [\partial_\alpha x_i \partial_\beta x_i] = \boldsymbol{F}\boldsymbol{F}^T, \\
\boldsymbol{g}^{-1} &= [g^{ij}] = [\partial_\alpha x_i \partial_\alpha x_j] = \boldsymbol{F}^T\boldsymbol{F}, \\
\boldsymbol{g} &= [g_{ij}] = [\partial_i X_\alpha \partial_j X_\alpha] = \boldsymbol{f}\boldsymbol{f}^T.
\end{aligned} \tag{5.29}$$

The reciprocal deformation tensors $G^{\alpha\beta}$ and g^{ij} are attributed to Piola and Finger, respectively. The characteristics of a deformed body in the neighborhood of a point $\boldsymbol{x}$ can be described as

$$\begin{aligned}
x_i(X_1 + dX_1, \ldots,) &= x_i(X_1, \ldots,) + dX_\alpha \partial_\alpha x_i(X_1, \ldots,) \\
&\quad + \frac{1}{2} dX_\alpha dX_\beta \partial_\alpha \partial_\beta x_i(X_1, \ldots,) + \cdots + .
\end{aligned}$$

In classical continuum mechanics it is postulated that the neighborhood can be as small as possible. Then, the second-order quantities $dX_\alpha dX_\beta$ render the third term negligible. As a result, the deformation gradients $\partial_\alpha x_i$ completely define the characteristics of the deformation. Materials satisfying this postulate are called **simple materials.** To take into account the coarse microstructure of certain materials, we are not permitted to assume that the third term is negligible, and we may have to include the second derivatives in the theories.

5.3 Strain Tensors

The lengths of elements before and after deformation are given by

$$(dS)^2 = dX_\alpha dX_\alpha = g_{ij} dx_i dx_j, \quad (ds)^2 = dx_i dx_i = G_{\alpha\beta} dX_\alpha dX_\beta. \tag{5.30}$$

The strain tensors $E_{\alpha\beta}$ and e_{ij} are measures of the change in length:

$$\begin{aligned}(ds)^2 - (dS)^2 &= 2E_{\alpha\beta} dX_\alpha dX_\beta = [G_{\alpha\beta} - \delta_{\alpha\beta}] dX_\alpha dX_\beta \\ &= 2e_{ij} dx_i dx_j = [\delta_{ij} - g_{ij}] dx_i dx_j.\end{aligned} \tag{5.31}$$

The strain tensor $\boldsymbol{E}$ is called the Green's strain, and $\boldsymbol{e}$ is called the Almansi strain. The names Lagrange strain and Cauchy strain are also used, respectively, for these two quantities. It is easy to see that both of these strains are symmetric tensors.

A relation between these two strain measures can be established as

$$e_{ij} = E_{\alpha\beta} \partial_i X_\alpha \partial_j X_\beta, \quad E_{\alpha\beta} = e_{ij} \partial_\alpha x_i \partial_\beta x_j, \tag{5.32}$$

which have the matrix forms

$$\boldsymbol{e} = \boldsymbol{f} \boldsymbol{E} \boldsymbol{f}^T, \quad \boldsymbol{E} = \boldsymbol{F} \boldsymbol{e} \boldsymbol{F}^T. \tag{5.33}$$

By use of

$$x_i = [X_\beta + U_\beta] \delta_{i\beta}, \quad X_\alpha = [x_j + u_j] \delta_{\alpha j}, \tag{5.34}$$

the deformation gradients become

$$\partial_\alpha x_i = \delta_{\alpha i} + \partial_\alpha U_\beta \delta_{i\beta}, \quad \partial_i X_\alpha = \delta_{i\alpha} - \partial_i u_j \delta_{\alpha j}. \tag{5.35}$$

The deformation tensors become

$$\begin{aligned}G_{\alpha\beta} &= \delta_{\alpha\beta} + \partial_\alpha U_\beta + \partial_\beta U_\alpha + \partial_\alpha U_\gamma \partial_\beta U_\gamma, \\ g_{ij} &= \delta_{ij} - \partial_i u_j - \partial_j u_i + \partial_i u_k \partial_j u_k.\end{aligned} \tag{5.36}$$

Now the strain tensors can be written as

$$\begin{aligned}E_{\alpha\beta} &= \frac{1}{2}[\partial_\alpha U_\beta + \partial_\beta U_\alpha + \partial_\alpha U_\gamma \partial_\beta U_\gamma], \\ e_{ij} &= \frac{1}{2}[\partial_i u_j + \partial_j u_i - \partial_i u_k \partial_j u_k].\end{aligned} \tag{5.37}$$

In engineering notation, using U, V, and W to represent U_α and X, Y, and Z to represent X_α, we have

$$
\begin{aligned}
E_{XX} &= \frac{\partial U}{\partial X} + \frac{1}{2}\left[\left(\frac{\partial U}{\partial X}\right)^2 + \left(\frac{\partial V}{\partial X}\right)^2 + \left(\frac{\partial W}{\partial X}\right)^2\right],\\
E_{YY} &= \frac{\partial V}{\partial Y} + \frac{1}{2}\left[\left(\frac{\partial U}{\partial Y}\right)^2 + \left(\frac{\partial V}{\partial Y}\right)^2 + \left(\frac{\partial W}{\partial Y}\right)^2\right],\\
E_{ZZ} &= \frac{\partial W}{\partial Z} + \frac{1}{2}\left[\left(\frac{\partial U}{\partial Z}\right)^2 + \left(\frac{\partial V}{\partial Z}\right)^2 + \left(\frac{\partial W}{\partial Z}\right)^2\right],\\
E_{XY} &= \frac{1}{2}\left[\frac{\partial U}{\partial Y} + \frac{\partial V}{\partial X} + \frac{\partial U}{\partial X}\frac{\partial U}{\partial Y} + \frac{\partial V}{\partial X}\frac{\partial V}{\partial Y} + \frac{\partial W}{\partial X}\frac{\partial W}{\partial Y}\right],\\
E_{YZ} &= \frac{1}{2}\left[\frac{\partial V}{\partial Z} + \frac{\partial W}{\partial Y} + \frac{\partial U}{\partial Y}\frac{\partial U}{\partial Z} + \frac{\partial V}{\partial Y}\frac{\partial V}{\partial Z} + \frac{\partial W}{\partial Y}\frac{\partial W}{\partial Z}\right],\\
E_{ZX} &= \frac{1}{2}\left[\frac{\partial W}{\partial X} + \frac{\partial U}{\partial Z} + \frac{\partial U}{\partial Z}\frac{\partial U}{\partial X} + \frac{\partial V}{\partial Z}\frac{\partial V}{\partial X} + \frac{\partial W}{\partial Z}\frac{\partial W}{\partial X}\right].
\end{aligned}
\tag{5.38}
$$

Similarly, the Almansi strain components can be written as

$$
\begin{aligned}
e_{xx} &= \frac{\partial u}{\partial x} - \frac{1}{2}\left[\left(\frac{\partial u}{\partial x}\right)^2 + \left(\frac{\partial v}{\partial x}\right)^2 + \left(\frac{\partial w}{\partial x}\right)^2\right],\\
e_{yy} &= \frac{\partial v}{\partial y} - \frac{1}{2}\left[\left(\frac{\partial u}{\partial y}\right)^2 + \left(\frac{\partial v}{\partial y}\right)^2 + \left(\frac{\partial w}{\partial y}\right)^2\right],\\
e_{zz} &= \frac{\partial w}{\partial z} - \frac{1}{2}\left[\left(\frac{\partial u}{\partial z}\right)^2 + \left(\frac{\partial v}{\partial z}\right)^2 + \left(\frac{\partial w}{\partial z}\right)^2\right],\\
e_{xy} &= \frac{1}{2}\left[\frac{\partial u}{\partial y} + \frac{\partial v}{\partial x} - \frac{\partial u}{\partial x}\frac{\partial u}{\partial y} - \frac{\partial v}{\partial x}\frac{\partial v}{\partial y} - \frac{\partial w}{\partial x}\frac{\partial w}{\partial y}\right],\\
e_{yz} &= \frac{1}{2}\left[\frac{\partial v}{\partial z} + \frac{\partial w}{\partial y} - \frac{\partial u}{\partial y}\frac{\partial u}{\partial z} - \frac{\partial v}{\partial y}\frac{\partial v}{\partial z} - \frac{\partial w}{\partial y}\frac{\partial w}{\partial z}\right],\\
e_{zx} &= \frac{1}{2}\left[\frac{\partial w}{\partial x} + \frac{\partial u}{\partial z} - \frac{\partial u}{\partial z}\frac{\partial u}{\partial x} - \frac{\partial v}{\partial z}\frac{\partial v}{\partial x} - \frac{\partial w}{\partial z}\frac{\partial w}{\partial x}\right].
\end{aligned}
\tag{5.39}
$$

5.3.1 Decomposition of Displacement Gradients

The displacement gradients $\partial_i u_j$ and $\partial_\alpha U_\beta$ can be resolved into symmetric and skew-symmetric tensors.

Let

$$\partial_i u_j = \bar{e}_{ij} + \bar{r}_{ij}, \quad \partial_\alpha U_\beta = \bar{E}_{\alpha\beta} + \bar{R}_{\alpha\beta}, \tag{5.40}$$

where $\bar{e}_{ij}$ and $\bar{E}_{\alpha\beta}$ are symmetric tensors called infinitesimal strains and $\bar{r}_{ij}$ and $\bar{R}_{\alpha\beta}$ are skew-symmetric tensors called infinitesimal rotations. These can be written as

$$\begin{aligned}
\bar{e}_{ij} &= \frac{1}{2}(\partial_i u_j + \partial_j u_i),\\
\bar{E}_{\alpha\beta} &= \frac{1}{2}(\partial_\alpha U_\beta + \partial_\beta U_\alpha),\\
\bar{r}_{ij} &= \frac{1}{2}(\partial_i u_j - \partial_j u_i),\\
\bar{R}_{\alpha\beta} &= \frac{1}{2}(\partial_\alpha U_\beta - \partial_\beta U_\alpha).
\end{aligned} \tag{5.41}$$

The Almansi strain components are related to the infinitesimal strains in the form

$$e_{ij} = \bar{e}_{ij} - \frac{1}{2}[\bar{e}_{ik} + \bar{r}_{ik}][\bar{e}_{jk} + \bar{r}_{jk}]. \tag{5.42}$$

When $|\bar{e}_{ij}|$ and $|\bar{r}_{ij}|$ are small, the Almansi strain is approximately equal to the infinitesimal strain. Further, as

$$E_{\alpha\beta} = e_{ij}\partial_\alpha x_i \partial_\beta x_j, \tag{5.43}$$

neglecting higher-order terms,

$$E_{11} = e_{11}, \ldots . \tag{5.44}$$

In other words, when the displacement gradients and rotations are small, the material description and spatial description are identical.

5.3.2 Stretch

Consider an element $d\boldsymbol{X}$ deforming into $d\boldsymbol{x}$. Let us attach unit vectors $\boldsymbol{N}$ to $d\boldsymbol{X}$ and $\boldsymbol{n}$ to $d\boldsymbol{x}$. Then,

$$\boldsymbol{N} = \frac{d\boldsymbol{X}}{dS}, \quad \boldsymbol{n} = \frac{d\boldsymbol{x}}{ds}. \tag{5.45}$$

In the component form,

$$N_\alpha = \frac{dX_\alpha}{dS}, \quad n_i = \frac{dx_i}{ds}. \tag{5.46}$$

Using

$$(ds)^2 = G_{\alpha\beta} dX_\alpha dX_\beta, \quad (dS)^2 = g_{ij} dx_i dx_j, \tag{5.47}$$

we get

$$\begin{aligned}
\left(\frac{ds}{dS}\right)^2 &= G_{\alpha\beta}\frac{dX_\alpha}{dS}\frac{dX_\beta}{dS} = G_{\alpha\beta} N_\alpha N_\beta,\\
\left(\frac{dS}{ds}\right)^2 &= g_{ij}\frac{dx_i}{ds}\frac{dx_j}{ds} = g_{ij} n_i n_j.
\end{aligned} \tag{5.48}$$

The ratio (ds/dS) is called the **stretch**. This ratio depends on the orientation of the element. We have

$$\Lambda_N = \frac{ds}{dS} = \sqrt{G_{\alpha\beta} N_\alpha N_\beta}, \quad \lambda_n = \frac{ds}{dS} = \frac{1}{\sqrt{g_{ij} n_i n_j}} \tag{5.49}$$

for the stretches in the material and spatial descriptions.

As special cases, if $\boldsymbol{N} = \boldsymbol{e}_1$, we get

$$N_1 = 1, \quad N_2 = 0, \quad N_3 = 0, \quad \Lambda_1 = \sqrt{G_{11}}, \tag{5.50}$$

and if $\boldsymbol{n} = \boldsymbol{e}_1$,

$$n_1 = 1, \quad n_2 = 0, \quad n_3 = 0, \quad \lambda_1 = \frac{1}{\sqrt{g_{11}}}. \tag{5.51}$$

(Note that $\Lambda_1 \neq \lambda_1$ as $\boldsymbol{N} = \boldsymbol{e}_1$ does not deform into $\boldsymbol{n} = \boldsymbol{e}_1$.) A similar consideration of unit normals in the other two directions yields the geometrical meaning of the diagonal components of the deformation tensors:

$$G_{11} = \Lambda_1^2, \quad G_{22} = \Lambda_2^2, \quad G_{33} = \Lambda_3^2. \tag{5.52}$$

To see the meaning of the off-diagonal elements, recall that $\boldsymbol{e}_\alpha$ in the undeformed body maps onto $\boldsymbol{G}_\alpha$ in the deformed body. If we take two orthogonal unit vectors $\boldsymbol{e}_1$ and $\boldsymbol{e}_2$, their images are $\boldsymbol{G}_1$ and $\boldsymbol{G}_2$. If θ_{12} denotes the angle between $\boldsymbol{G}_1$ and $\boldsymbol{G}_2$, we have

$$\cos\theta_{12} = \frac{\boldsymbol{G}_1 \cdot \boldsymbol{G}_2}{|\boldsymbol{G}_1||\boldsymbol{G}_2|} = \frac{G_{12}}{\sqrt{G_{11}G_{22}}}. \tag{5.53}$$

Thus G_{12} indicates the distortion of the original angle of 90°.

5.3.3 Extension

The extensions E_N and e_n are defined as

$$E_N = \Lambda_N - 1, \quad e_n = \lambda_n - 1. \tag{5.54}$$

Extension, stretch, deformation, and strain are related in the form

$$\begin{aligned} E_1 &= \Lambda_1 - 1 = \sqrt{G_{11}} - 1 = \sqrt{2E_{11} + 1} - 1, \\ e_1 &= \lambda_1 - 1 = \frac{1}{\sqrt{g_{11}}} - 1 = \frac{1}{\sqrt{1 - 2e_{11}}} - 1. \end{aligned} \tag{5.55}$$

5.3.4 Infinitesimal Strains and Rotations

When the difference between ds and dS is small, for an element initially in the 1-direction, we find

$$\left(\frac{ds}{dS}\right)^2 - 1 = \frac{ds - dS}{dS}\frac{ds + dS}{dS} = 2E_{11}. \tag{5.56}$$

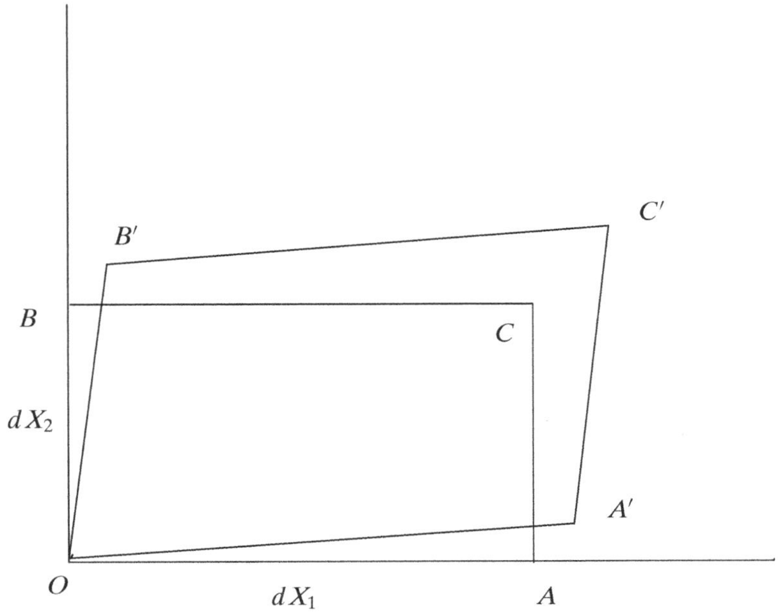

Figure 5.4. Infinitesimal deformation of a rectangle.

Using

$$\frac{ds + dS}{dS} \approx 2, \tag{5.57}$$

we get

$$E_{11} = \frac{ds - dS}{dS}, \tag{5.58}$$

which is the elementary definition of strain as the ratio of change in length to the original length.

Consider a rectangular material element $OACB$ with O at (X_1, X_2) and $OA = dX_1$ and $OB = dX_2$, as shown in Fig. 5.4. During deformation, O moves by (U_1, U_2) to O' and A, B, and C end up at A', B', and C'. Using a rigid body translation, we can move the element back to make O' coincide with O. Then the relative coordinates of the other points are

$$\begin{aligned} A' &: (dX_1 + \partial_1 U_1 dX_1, \partial_1 U_2 dX_1), \\ B' &: (\partial_2 U_1 dX_2, dX_2 + \partial_2 U_2 dX_2), \\ C' &: (dX_1 + \partial_1 U_1 dX_1 + \partial_2 U_1 dX_2, dX_2 + \partial_1 U_2 dX_1 + \partial_2 U_2 dX_2). \end{aligned}$$

Assuming small displacement derivatives, we find

$$E_{11} = \partial_1 U_1, \quad E_{22} = \partial_2 U_2. \tag{5.59}$$

The counterclockwise rotations θ_{12} of OA and θ_{21} of OB are obtained as

$$\theta_{12} = \partial_1 U_2, \quad \theta_{21} = -\partial_2 U_1. \tag{5.60}$$

The engineering shear strain Γ_{12} and the mathematical shear strain E_{12} are measures of the change in angle from the original 90°:

$$\Gamma_{12} = 2E_{12} = \partial_2 U_1 + \partial_1 U_2. \tag{5.61}$$

The infinitesimal rotation R_{12} is defined as the average of the two rotations,

$$R_{12} = \frac{1}{2}[\partial_1 U_2 - \partial_2 U_1]. \tag{5.62}$$

It turns out that, if we consider an element with initial arbitrary orientation θ and final orientation $\theta + \phi$, the infinitesimal rotation R_{12} is the average of $\tan\phi$, with averaging done over all values of θ's.

5.3.5 Deformation Ellipsoids

Consider a small sphere in the undeformed configuration $dX_\alpha dX_\alpha = K^2$, with the center at $\boldsymbol{X}$. In the deformed configuration, the center has moved to $\boldsymbol{x}$. The map of the sphere can be found from

$$dX_\alpha dX_\alpha = \partial_i X_\alpha \partial_j X_\alpha dx_i dx_j = g_{ij} dx_i dx_j = K^2. \tag{5.63}$$

As g_{ij} is positive definite, the preceding quadratic form represents an ellipsoid, known as the material ellipsoid of Cauchy (see Fig. 5.5).

Similarly, a sphere $dx_i dx_i = k^2$ in the deformed configuration is the map of an ellipsoid,

$$dx_i dx_i = \partial_\alpha x_i \partial_\beta x_i dX_\alpha dX_\beta = G_{\alpha\beta} dX_\alpha dX_\beta = k^2, \tag{5.64}$$

in the undeformed configuration. This is known as the spatial ellipsoid of Cauchy (Fig. 5.5).

Let us concentrate on the spatial ellipsoid for the time being. The directions of the three axes of the ellipsoid correspond to the direction along which the stretch

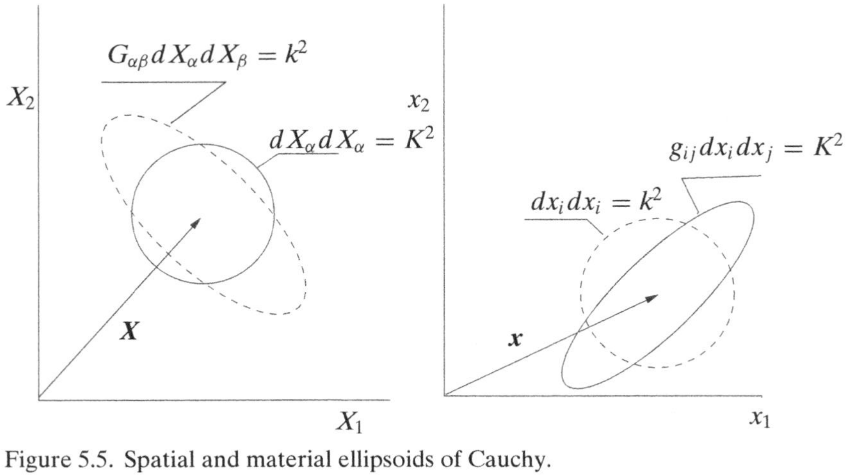

Figure 5.5. Spatial and material ellipsoids of Cauchy.

$\Lambda = ds/dS$ is stationary. If $\boldsymbol{N}$ is a unit vector pointing in the direction of $d\boldsymbol{X}$ inside the ellipsoid in the undeformed configuration, we have

$$N_\alpha = \frac{dX_\alpha}{dS}. \tag{5.65}$$

By dividing Eq. (5.64) by $(dS)^2$, we find

$$G_{\alpha\beta} N_\alpha N_\beta = \Lambda_N^2. \tag{5.66}$$

To find the directions $\boldsymbol{N}$ along which Λ_N^2 is stationary, we have to extremize it subject to the condition that $\boldsymbol{N}$ is a unit vector. For such constrained extremization, as discussed in Chapter 2, we use the method of Lagrange multipliers and define a new function,

$$F(N_1, N_2, N_3) = G_{\alpha\beta} N_\alpha N_\beta - G(N_\alpha N_\alpha - 1), \tag{5.67}$$

where G is the Lagrange multiplier. Setting

$$\frac{\partial F}{\partial N_\alpha} = 0, \tag{5.68}$$

we obtain a system of simultaneous equations,

$$G_{\alpha\beta} N_\beta - G N_\alpha = 0, \tag{5.69}$$

where we have used the fact that $G_{\alpha\beta}$ is symmetric.

In expanded form, we have

$$\begin{bmatrix} G_{11} - G & G_{12} & G_{13} \\ G_{12} & G_{22} - G & G_{23} \\ G_{13} & G_{23} & G_{33} - G \end{bmatrix} \begin{Bmatrix} N_1 \\ N_2 \\ N_3 \end{Bmatrix} = \begin{Bmatrix} 0 \\ 0 \\ 0 \end{Bmatrix}. \tag{5.70}$$

This is a typical matrix eigenvalue problem. For a nontrivial solution $\boldsymbol{N}$, we must have

$$\det(\boldsymbol{G} - G\boldsymbol{I}) = 0. \tag{5.71}$$

We may expand the determinant to obtain the characteristic equation of the matrix $\boldsymbol{G}$ in the form

$$-G^3 + I_{G1} G^2 - I_{G2} G + I_{G3} = 0, \tag{5.72}$$

where the coefficients of the cubic are the three invariants of the matrix $\boldsymbol{G}$, namely,

$$\begin{aligned} I_{G1} &= G_{11} + G_{22} + G_{33}, \\ I_{G2} &= G_{22}G_{33} + G_{33}G_{11} + G_{11}G_{22} - G_{23}^2 - G_{31}^2 - G_{12}^2 \\ &= \begin{vmatrix} G_{22} & G_{23} \\ G_{32} & G_{33} \end{vmatrix} + \begin{vmatrix} G_{11} & G_{31} \\ G_{13} & G_{33} \end{vmatrix} + \begin{vmatrix} G_{11} & G_{12} \\ G_{21} & G_{22} \end{vmatrix}, \\ I_{G3} &= \begin{vmatrix} G_{11} & G_{12} & G_{13} \\ G_{21} & G_{22} & G_{23} \\ G_{31} & G_{32} & G_{33} \end{vmatrix}. \end{aligned} \tag{5.73}$$

Let $G^{(\alpha)}$ denote the three roots of cubic (5.72). They satisfy

$$
\begin{aligned}
I_{G1} &= G^{(1)} + G^{(2)} + G^{(3)}, \\
I_{G2} &= G^{(1)}G^{(2)} + G^{(2)}G^{(3)} + G^{(3)}G^{(1)}, \\
I_{G3} &= G^{(1)}G^{(2)}G^{(3)}.
\end{aligned} \tag{5.74}
$$

Substituting $G = G^{(\gamma)}$ in Eq. (5.70) for each value of γ, we find a vector $\boldsymbol{N}^{(\gamma)}$ (after normalizing it to have unit length). This gives us three directions along which Λ_N is stationary. These are called the principal axes of the ellipsoid. For a symmetric matrix, it is known that these three eigenvectors are mutually orthogonal (or when the cubic has repeated roots, the directions can be selected to be orthogonal) and the eigenvalues $G^{(\gamma)}$ are real.

$$
\boldsymbol{N}^{(\alpha)} \cdot \boldsymbol{N}^{(\beta)} = \delta_{\alpha\beta}. \tag{5.75}
$$

Multiplying $G_{\alpha\beta}N_\beta^{(\gamma)} = GN_\alpha^{(\gamma)}$ by $N_\alpha^{(\gamma)}$ (summation on α only), we obtain

$$
G^{(\gamma)} = G_{\alpha\beta}N_\alpha^{(\gamma)}N_\beta^{(\gamma)} = \Lambda_{(\gamma)}^2. \tag{5.76}
$$

The eigenvalues happen to be the squares of the principal stretches. The deformation tensor $\boldsymbol{G}$ has the spectral representation

$$
\boldsymbol{G} = \sum_{\gamma=1}^{3} \Lambda_{(\gamma)}^2 \boldsymbol{N}^{(\gamma)}\boldsymbol{N}^{(\gamma)T}. \tag{5.77}
$$

The analysis of the material ellipsoid $g_{ij}dx_i dx_j = K^2$ can be carried out in a similar manner. The eigenvalues of $\boldsymbol{g}$ will be denoted by $1/\lambda_{(i)}^2$ and the eigenvectors by $\boldsymbol{n}^{(i)}$. It has the spectral representation

$$
\boldsymbol{g} = \sum_{i=1}^{3} \frac{1}{\lambda_{(i)}^2} \boldsymbol{n}^{(i)}\boldsymbol{n}^{(i)T}. \tag{5.78}
$$

The extremum values of the strain tensors $\boldsymbol{E}$ and $\boldsymbol{e}$ and their directions (principal strains and principal directions) are related to the principal stretches and their directions in a simple fashion.

In the eigenvalue problem,

$$
G_{\alpha\beta}N_\beta^{(\gamma)} - G^{(\gamma)}N_\alpha^{(\gamma)} = 0, \tag{5.79}
$$

if we substitute

$$
G_{\alpha\beta} = 2E_{\alpha\beta} + \delta_{\alpha\beta}, \tag{5.80}
$$

we get

$$
E_{\alpha\beta}N_\beta^{(\gamma)} - \frac{1}{2}(G^{(\gamma)} - 1)N_\alpha^{(\gamma)} = 0. \tag{5.81}
$$

The principal directions of the strain tensor $\boldsymbol{E}$ are the same as $\boldsymbol{N}^{(\gamma)}$, and the principal strains are

$$E^{(\gamma)} = \frac{1}{2}(G^{(\gamma)} - 1). \tag{5.82}$$

Similarly,

$$e^{(i)} = \frac{1}{2}(1 - g^{(i)}). \tag{5.83}$$

Thus the solution of the eigenvalue problem for $\boldsymbol{G}$ or $\boldsymbol{g}$ also gives the principal strains and their directions. We note that the strain matrix is not positive definite and the term "strain ellipsoid" will be incorrect.

5.3.6 Polar Decomposition of the Deformation Gradient

The deformation gradient tensor

$$F_{\alpha i} = \partial_\alpha x_i \tag{5.84}$$

has the property

$$\boldsymbol{F}\boldsymbol{F}^T = \boldsymbol{G}, \tag{5.85}$$

where, as we have seen, $\boldsymbol{G}$ is symmetric and positive definite, but $\boldsymbol{F}$ is neither.

From the theory of matrices, we know that we can factor any real, square matrix into an orthogonal matrix and a positive-definite symmetric matrix. The deformation gradient matrix $\boldsymbol{F}$ can be expressed as

$$\boldsymbol{F} = \boldsymbol{R}\boldsymbol{U} = \boldsymbol{V}\boldsymbol{R}, \tag{5.86}$$

where $\boldsymbol{R}$ is an orthogonal matrix (rotation matrix) and $\boldsymbol{U}$ is a positive-definite symmetric matrix known as the right-stretch matrix. The matrix $\boldsymbol{V}$ is called the left-stretch matrix. This factorization is called the polar decomposition. As discussed in Chapter 2, the orthogonal matrix $\boldsymbol{R}$ satisfies

$$\boldsymbol{R}^T\boldsymbol{R} = \boldsymbol{I}, \quad \boldsymbol{R}\boldsymbol{R}^T = \boldsymbol{I}, \quad \boldsymbol{R}^T = \boldsymbol{R}^{-1}. \tag{5.87}$$

Using these properties, from Eq. (5.86), we get

$$\begin{aligned} \boldsymbol{F}^T\boldsymbol{F} &= \boldsymbol{U}^2 = [g^{ij}] \;\; = \left[\partial_\alpha x^i \partial_\alpha x^j\right], \\ \boldsymbol{F}\boldsymbol{F}^T &= \boldsymbol{V}^2 = [G_{\alpha\beta}] = \left[\partial_\alpha x^i \partial_\beta x^i\right]. \end{aligned}$$

We note that the eigenvectors $\boldsymbol{N}^{(\gamma)}$ of $\boldsymbol{G}$ are the same as those of $\boldsymbol{V}$. The spectral representation of $\boldsymbol{V}$ is

$$\boldsymbol{V} = \sum_{\gamma=1}^{3} \Lambda_{(\gamma)} \boldsymbol{N}^{(\gamma)} \boldsymbol{N}^{(\gamma)T}. \tag{5.88}$$

In a similar way, we may resolve the Cauchy deformation gradient $\boldsymbol{f}$ as

$$\boldsymbol{f} = \boldsymbol{r}\boldsymbol{u} = \boldsymbol{v}\boldsymbol{r}, \tag{5.89}$$

where

$$\boldsymbol{r}^T\boldsymbol{r} = \boldsymbol{r}\boldsymbol{r}^T = \boldsymbol{I}, \tag{5.90}$$

and

$$\begin{aligned}\boldsymbol{f}^T\boldsymbol{f} &= \boldsymbol{u}^2 = [G^{\alpha\beta}] = \left[\partial_i X^\alpha \partial_i X^\beta\right],\\ \boldsymbol{f}\boldsymbol{f}^T &= \boldsymbol{v}^2 = [g_{ij}] \;\; = [\partial_i X^\alpha \partial_j X^\alpha].\end{aligned}$$

From $\boldsymbol{F}\boldsymbol{f} = \boldsymbol{I}$,

$$\boldsymbol{V}\boldsymbol{u} = \boldsymbol{I}, \quad \boldsymbol{U}\boldsymbol{v} = \boldsymbol{I}, \quad \boldsymbol{R}\boldsymbol{r} = \boldsymbol{I}. \tag{5.91}$$

In terms of the eigenvalues and eigenvectors of $\boldsymbol{g}$, $1/\lambda_{(i)}$, and $\boldsymbol{n}^{(i)}$,

$$\boldsymbol{v} = \sum_{i=1}^{3} \frac{1}{\lambda_{(i)}} \boldsymbol{n}^{(i)} \boldsymbol{n}^{(i)T}. \tag{5.92}$$

5.3.7 Stretch and Rotation

In the relations

$$d\boldsymbol{x} = \boldsymbol{F}^T d\boldsymbol{X}, \quad d\boldsymbol{X} = \boldsymbol{f}^T d\boldsymbol{x}, \tag{5.93}$$

using the polar decomposition, we get

$$d\boldsymbol{x} = \boldsymbol{U}\boldsymbol{R}^T d\boldsymbol{X} = \boldsymbol{R}^T \boldsymbol{V} d\boldsymbol{X}, \quad d\boldsymbol{X} = \boldsymbol{u}\boldsymbol{r}^T d\boldsymbol{x} = \boldsymbol{r}^T \boldsymbol{v} d\boldsymbol{x} \tag{5.94}$$

The operations on $d\boldsymbol{X}$ show that, in the first case, the material element is rotated by $\boldsymbol{R}^T$ and then stretched by $\boldsymbol{U}$, and in the second case, it is first stretched by $\boldsymbol{V}$ and then rotated by $\boldsymbol{R}^T$.

For an arbitrary unit vector $\boldsymbol{N} = d\boldsymbol{X}/dS$, we obtain

$$d\boldsymbol{x} = \boldsymbol{R}^T \boldsymbol{V}\boldsymbol{N} dS.$$

Writing $\boldsymbol{n} = d\boldsymbol{x}/ds$, we have

$$\boldsymbol{n} = \frac{dS}{ds} \boldsymbol{R}^T \boldsymbol{V}\boldsymbol{N}. \tag{5.95}$$

If $\boldsymbol{N} = \boldsymbol{N}^{(\gamma)}$, a principal direction of $\boldsymbol{G}$, $\boldsymbol{V}\boldsymbol{N}^{(\gamma)} = \Lambda_{(\gamma)}\boldsymbol{N}^{(\gamma)}$ and

$$\boldsymbol{n} = \boldsymbol{R}^T \boldsymbol{N}^{(\gamma)}. \tag{5.96}$$

Thus the unit vector along a principal direction undergoes a rigid body rotation. A triad of mutually orthogonal unit vectors representing the three principal directions will get transformed to a new triad of orthogonal unit vectors.

Using $\boldsymbol{f} = \boldsymbol{R}^T \boldsymbol{V}^{-1}$, we obtain

$$\boldsymbol{g} = \boldsymbol{R}^T \boldsymbol{V}^{-2} \boldsymbol{R}, \tag{5.97}$$

$$\boldsymbol{g}\boldsymbol{n} = \boldsymbol{R}^T \boldsymbol{V}^{-2} \boldsymbol{R}\boldsymbol{R}^T \boldsymbol{N}^{(\gamma)} = \frac{1}{\Lambda_{(\gamma)}^2} \boldsymbol{R}^T \boldsymbol{N}^{(\gamma)} = \frac{\boldsymbol{n}}{\Lambda_{(\gamma)}^2}. \tag{5.98}$$

Thus $\boldsymbol{n}$ is an eigenvector of $\boldsymbol{g}$, and the new triad just mentioned points along the principal directions of the material ellipsoid of Cauchy. Also, the largest semimajor axis of the spatial ellipsoid gets mapped into the smallest semimajor axis of the material ellipsoid.

To find the factors $\boldsymbol{U}$ and $\boldsymbol{V}$, we need the square root of the positive-definite matrix $\boldsymbol{G}$. We can use the Cayley–Hamilton theorem, which states that any square matrix satisfies its own characteristic equation, to obtain the square root of a matrix. The characteristic equation for $\boldsymbol{G}$ is Eq. (5.72), and

$$P(G) \equiv -G^3 + I_{G1}G^2 - I_{G2}G + I_{G3} = 0. \tag{5.99}$$

According to the Cayley–Hamilton theorem, the matrix $\boldsymbol{G}$ satisfies the matrix polynomial equation

$$\boldsymbol{P}(\boldsymbol{G}) \equiv -\boldsymbol{G}^3 + I_{G1}\boldsymbol{G}^2 - I_{G2}\boldsymbol{G} + I_{G3}\boldsymbol{I} = \boldsymbol{O}, \tag{5.100}$$

where $\boldsymbol{O}$ is a 3×3 matrix with all elements zero. We have also used the notation $\boldsymbol{P}$ for the matrix function obtained from the regular function P. This result implies that any power of $\boldsymbol{G}$ higher than 2 can be expressed in terms of $\boldsymbol{G}^2$, $\boldsymbol{G}$, and $\boldsymbol{I}$. In fact, any matrix function $\boldsymbol{\phi}$ of $\boldsymbol{G}$ can be expressed the same way. Moreover, the eigenvectors of $\boldsymbol{G}$ are also the eigenvectors of $\boldsymbol{\phi}$, and the eigenvalues of $\boldsymbol{\phi}(\boldsymbol{G})$ are $\phi(G)$.

Thus

$$\boldsymbol{G}^{1/2} = a\boldsymbol{G}^2 + b\boldsymbol{G} + c\boldsymbol{I}, \tag{5.101}$$

where the coefficients a, b, and c have to be found.

For the three eigenvalues $G^{(\gamma)}$ of $\boldsymbol{G}$, the eigenvalues of $\boldsymbol{G}^{1/2}$ are $(G^{(\gamma)})^{1/2}$. Using these, we get three equations,

$$(G^{(\gamma)})^{1/2} = a(G^{(\gamma)})^2 + bG^{(\gamma)} + c, \quad \gamma = 1, 2, 3, \tag{5.102}$$

for a, b, and c.

When there are repeated roots $G^{(\gamma)}$, we may use the Frobenius method, in which we form a new equation for a, b, c by differentiating Eq. (5.101) with respect to G.

5.3.8 Example: Polar Decomposition

Consider the deformation of a square region in the X_1, X_2 plane. First, let us stretch this region in the X_1 direction, doubling the length. Then, we rotate it by $\theta = 30°$ (see Fig. 5.6). The spatial coordinates of an initial point X_1, X_2 become

$$x_1 = 2X_1 \cos\theta - X_2 \sin\theta,$$
$$x_2 = 2X_1 \sin\theta + X_2 \cos\theta. \tag{5.103}$$

The Green's deformation gradient matrix is obtained as

$$\boldsymbol{F} = \begin{bmatrix} \sqrt{3} & 1 \\ -1/2 & \sqrt{3}/2 \end{bmatrix}. \tag{5.104}$$

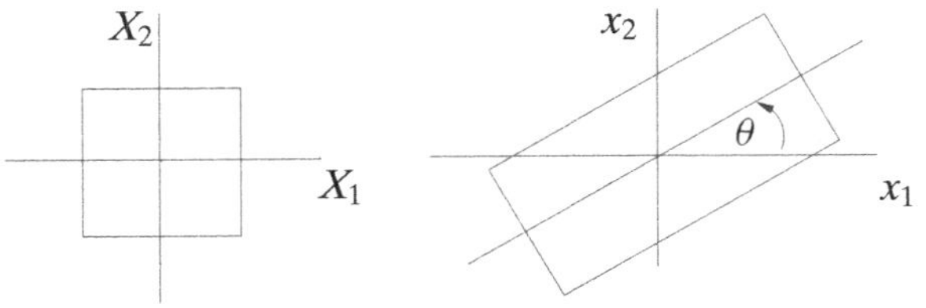

Figure 5.6. Stretching and rotation.

The Green's deformation matrix is obtained as

$$\boldsymbol{G} = \boldsymbol{F}\boldsymbol{F}^T = \begin{bmatrix} 4 & 0 \\ 0 & 1 \end{bmatrix}. \tag{5.105}$$

The characteristic equation for $\boldsymbol{G}$ is found as

$$\begin{vmatrix} 4 - G & 0 \\ 0 & 1 - G \end{vmatrix} = 0. \tag{5.106}$$

The eigenvalues are

$$G^{(1)} = 4, \quad G^{(2)} = 1. \tag{5.107}$$

The corresponding eigenvectors are

$$\boldsymbol{N}^{(1)} = \begin{Bmatrix} 1 \\ 0 \end{Bmatrix}, \quad \boldsymbol{N}^{(2)} = \begin{Bmatrix} 0 \\ 1 \end{Bmatrix}. \tag{5.108}$$

The spatial ellipse for this plane deformation is given by

$$\left(\frac{dX_1}{0.5}\right)^2 + \left(\frac{dX_2}{1}\right)^2 = 1, \tag{5.109}$$

where we have assumed the radius of the spatial sphere is $k = 1$. The square root of the diagonal matrix $\boldsymbol{G}$ is

$$\boldsymbol{V} = \begin{bmatrix} 2 & 0 \\ 0 & 1 \end{bmatrix}, \tag{5.110}$$

and, using this, we obtain

$$\boldsymbol{R} = \begin{bmatrix} \frac{\sqrt{3}}{2} & \frac{1}{2} \\ -\frac{1}{2} & \frac{\sqrt{3}}{2} \end{bmatrix}. \tag{5.111}$$

This corresponds to the 30° rotation we imparted to our region.

By inverting the matrix $\boldsymbol{F}$, we get

$$\boldsymbol{f} = \begin{bmatrix} \sqrt{3}/4 & -1/2 \\ 1/4 & \sqrt{3}/2 \end{bmatrix}, \tag{5.112}$$

$$\boldsymbol{g} = \boldsymbol{f}\boldsymbol{f}^T = \frac{1}{16}\begin{bmatrix} 7 & -3\sqrt{3} \\ -3\sqrt{3} & 13 \end{bmatrix}. \tag{5.113}$$

The eigenvalues of $\boldsymbol{g}$ are

$$\frac{1}{\lambda_{(1)}^2} = \frac{1}{4}, \quad \frac{1}{\lambda_{(2)}^2} = 1, \tag{5.114}$$

and the corresponding principal directions are

$$\boldsymbol{n}^{(1)} = \begin{Bmatrix} \frac{\sqrt{3}}{2} \\ \frac{1}{2} \end{Bmatrix}, \quad \boldsymbol{n}^{(2)} = \begin{Bmatrix} -\frac{1}{2} \\ \frac{\sqrt{3}}{2} \end{Bmatrix}. \tag{5.115}$$

A circle in the undeformed configuration,

$$(dX_1)^2 + (dX^2) = 1, \tag{5.116}$$

gets mapped into the ellipse,

$$\frac{7}{16}dx_1{}^2 - \frac{6\sqrt{3}}{16}dx_1 dx_2 + \frac{13}{16}(dx_2)^2 = 1, \tag{5.117}$$

which has a semimajor axis of 2 and semiminor axis of 1.

5.3.9 Example: Square Root of a Matrix

Consider the deformation gradient matrix

$$\boldsymbol{F} = \begin{bmatrix} \sqrt{3} & 1 & 0 \\ 0 & 2 & 0 \\ 0 & 0 & 1 \end{bmatrix}. \tag{5.118}$$

The Green's deformation matrix is obtained as

$$\boldsymbol{G} = \boldsymbol{F}\boldsymbol{F}^T = \begin{bmatrix} 4 & 2 & 0 \\ 2 & 4 & 0 \\ 0 & 0 & 1 \end{bmatrix}. \tag{5.119}$$

The eigenvalues are obtained from

$$[(4 - G)^2 - 4](1 - G) = 0, \tag{5.120}$$

which has the roots

$$G^{(1)} = 1, \quad G^{(2)} = 2, \quad G^{(3)} = 6. \tag{5.121}$$

Writing

$$\boldsymbol{V} = \boldsymbol{G}^{1/2} = a\boldsymbol{G}^2 + b\boldsymbol{G} + c\boldsymbol{I}, \tag{5.122}$$

we find the equations for a, b, and c as

$$\begin{aligned} 1^2 a + 1b + c &= \sqrt{1}, \\ 2^2 a + 2b + c &= \sqrt{2}, \\ 6^2 a + 6b + c &= \sqrt{6}. \end{aligned} \tag{5.123}$$

Solution of these equations gives

$$a = \frac{4 - 5\sqrt{2} + \sqrt{6}}{20}, \quad b = \frac{-32 + 35\sqrt{2} - 3\sqrt{6}}{20}, \quad c = \frac{48 - 30\sqrt{2} + 2\sqrt{6}}{20}. \tag{5.124}$$

Using these, we find

$$\boldsymbol{V} = \boldsymbol{G}^{1/2} = \frac{1}{\sqrt{2}} \begin{bmatrix} \sqrt{3}+1 & \sqrt{3}-1 & 0 \\ \sqrt{3}-1 & \sqrt{3}+1 & 0 \\ 0 & 0 & \sqrt{2} \end{bmatrix}. \tag{5.125}$$

The rotation matrix $\boldsymbol{R}$ and the left-stretch matrix $\boldsymbol{U}$ are found as

$$\boldsymbol{R} = \boldsymbol{V}^{-1}\boldsymbol{F} = \frac{1}{2\sqrt{6}} \begin{bmatrix} \sqrt{3}+1 & \sqrt{3}-1 & 0 \\ 1-\sqrt{3} & \sqrt{3}+1 & 0 \\ 0 & 0 & 2\sqrt{6} \end{bmatrix}, \tag{5.126}$$

$$\boldsymbol{U} = \boldsymbol{R}^T\boldsymbol{F} = \frac{1}{2\sqrt{2}} \begin{bmatrix} 3+\sqrt{3} & 3-\sqrt{3} & 0 \\ 3-\sqrt{3} & 3\sqrt{3}+1 & 0 \\ 0 & 0 & 2\sqrt{2} \end{bmatrix}. \tag{5.127}$$

Of course, it is easy to see in the preceding calculation that we dealt with only a 2×2 submatrix!

5.4 Logarithmic Strain

Apart from the Green's and Almansi strain tensors, a third strain measure frequently used is the logarithmic strain. If we consider a bar of current length L, stretched by ΔL, the incremental strain is

$$\Delta E^L = \frac{\Delta L}{L}, \tag{5.128}$$

which has the infinitesimal form

$$dE^L = \frac{dL}{L}. \tag{5.129}$$

Integrating this, we find, when the length has changed from L_0 to L,

$$E^L = \ln(L/L_0). \tag{5.130}$$

Using the left-stretch tensor $\boldsymbol{V}$, we have the tensor version of this relation:

$$\boldsymbol{E}^L = \ln \boldsymbol{V}. \tag{5.131}$$

5.5 Change of Volume

Let us consider a small parallelepiped formed by three material elements, $d\boldsymbol{X}$, $d\boldsymbol{X}^*$, and $d\boldsymbol{X}^{**}$. The volume enclosed is the scalar product of these elements:

$$dV = d\boldsymbol{X} \cdot d\boldsymbol{X}^* \times d\boldsymbol{X}^{**} = e_{\alpha\beta\gamma} dX_\alpha dX^*_\beta dX^{**}_\gamma. \tag{5.132}$$

Under deformation, these elements map onto $\boldsymbol{dx}$, $\boldsymbol{dx}^*$, and $\boldsymbol{dx}^{**}$, respectively. The new volume is

$$\begin{aligned} dv &= \boldsymbol{dx} \cdot \boldsymbol{dx}^* \times \boldsymbol{dx}^{**} = e_{ijk} dx_i dx_j^* dx_k^{**} \\ &= e_{ijk} \partial_\alpha x_i \partial_\beta x_j \partial_\gamma x_k dX_\alpha dX_\beta^* dX_\gamma^{**} \\ &= e_{\alpha\beta\gamma} dX_\alpha dX_\beta^* dX_\gamma^{**} \det[F] \\ &= JdV, \end{aligned} \tag{5.133}$$

where we have used

$$e_{ijk} \partial_\alpha x_i \partial_\beta x_j \partial_\gamma x_k = e_{\alpha\beta\gamma} J. \tag{5.134}$$

Using $j = 1/J$, we obtain

$$dV = jdv. \tag{5.135}$$

If we multiply this by $\partial_m X_\alpha$ and use the relation

$$\partial_m X_\alpha \partial_\alpha x_i = \delta_{im}, \tag{5.136}$$

we get an expression for the cofactors in a determinant:

$$e_{ijk} \partial_\beta x_j \partial_\gamma x_k = e_{\alpha\beta\gamma} J \partial_i X_\alpha. \tag{5.137}$$

5.6 Change of Area

Consider an element of area dA in the undeformed configuration, with a unit normal $\boldsymbol{N}$. Vectorial representation of this area element is

$$d\boldsymbol{A} = \boldsymbol{N} dA. \tag{5.138}$$

In this representation, the direction of $\boldsymbol{N}$ is not unique. However, if $\boldsymbol{A}$ is viewed as a vector product of two material elements $d\boldsymbol{X}$ and $d\boldsymbol{X}^*$, in the form

$$\begin{aligned} d\boldsymbol{A} &= d\boldsymbol{X} \times d\boldsymbol{X}^* = e_{\alpha\beta\gamma} dX_\beta dX_\gamma^* \boldsymbol{e}_\alpha, \\ dA_\alpha &= e_{\alpha\beta\gamma} dX_\beta dX_\gamma^*, \end{aligned} \tag{5.139}$$

the direction of $\boldsymbol{N}$ is unique.

Deformation of the body maps $d\boldsymbol{X}$ onto $\boldsymbol{dx}$ and $d\boldsymbol{X}^*$ onto $\boldsymbol{dx}^*$, and the deformed elements form the area

$$\begin{aligned} d\boldsymbol{a} &= \boldsymbol{n} da = \boldsymbol{dx} \times \boldsymbol{dx}^*, \\ da_i &= dx_j dx_k^* e_{ijk} \\ &= \partial_\beta x_j \partial_\gamma x_k \boldsymbol{e}_i dX_\beta dX_\gamma^*. \end{aligned} \tag{5.140}$$

Using the **cofactor** expansion, Eq. (5.137), we get

$$da_i = J \partial_i X_\alpha dA_\alpha. \tag{5.141}$$

This relation has as its inverse

$$dA_\alpha = j\partial_\alpha x_i da_i. \qquad (5.142)$$

5.7 Compatibility Equations

In three dimensions the strain tensor $E_{\alpha\beta}$ has six components that are related to the three displacement components U_α in the form

$$2E_{\alpha\beta} = G_{\alpha\beta} - \delta_{\alpha\beta} = \partial_\alpha U_\beta + \partial_\beta U_\alpha + \partial_\alpha U_\gamma \partial_\beta U_\gamma. \qquad (5.143)$$

If the displacements are known, it is straightforward to get the strain components through differentiation. If we reverse this situation and assume that six functions are given for the strain components, we have to integrate the preceding six equations to find three displacements. This poses a problem of an overdetermined system. The given six strain functions have to be constrained in some way to guarantee single-valued continuous displacements. These required restrictions are called the **compatibility** equations. A crude incompatible strain distribution is shown in Fig. 5.7. Here, the undeformed body has been cut into small pieces. Prescribing a strain distribution is the same as stretching and distorting each piece as we wish. Next, we try to assemble them into the deformed body. The arbitrary stretches and distortions result in a body with holes and a shape that depends on the order of assembling the pieces.

In the case of infinitesimal strains, we have

$$2E_{\alpha\beta} = G_{\alpha\beta} - \delta_{\alpha\beta} = \partial_\alpha U_\beta + \partial_\beta U_\alpha. \qquad (5.144)$$

From these we can eliminate U_α to obtain six relations among $E_{\alpha\beta}$. A compact way to write these six equations is

$$\nabla \times (\nabla \times \boldsymbol{E})^T = \boldsymbol{O}. \qquad (5.145)$$

In two dimensions, there is a single relation:

$$E_{11,22} + E_{22,11} = 2E_{12,12}. \qquad (5.146)$$

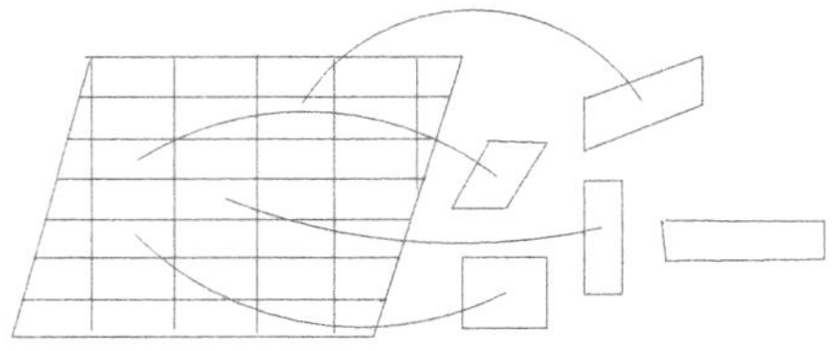

Figure 5.7. Incompatibility of arbitrary strains.

For the case of finite strains, the elimination method is very cumbersome, if not impossible. An alternative approach is to use the fact that the deformed body exists in the flat three-dimensional (3D) space; that is, the Riemann–Christoffel curvature has to be zero. With $\boldsymbol{G}_\alpha$ as base vectors and $G_{\alpha\beta}$ as the metric tensor, we can set

$$R_{\alpha\beta\gamma\delta} = 0. \tag{5.147}$$

Recall the expression for $\boldsymbol{R}$ from Chapter 3. It turns out, if we linearize the curvature expression, we get back the six compatibility equations previously mentioned.

5.8 Spatial Rotation and Two-Point Tensors

So far we have used a single Cartesian coordinate system, and if we rotate this system, both X_α and x_i change according to

$$X'_\alpha = Q_{\alpha\beta} X_\beta, \quad x'_i = Q_{ij} x_j, \tag{5.148}$$

where $\boldsymbol{Q}$ is a rotation matrix. Quantities such as $G_{\alpha\beta}$, $F_{\alpha i}$, and g_{ij} transform as

$$\boldsymbol{G}' = \boldsymbol{Q}\boldsymbol{G}\boldsymbol{Q}^T, \quad \boldsymbol{F}' = \boldsymbol{Q}\boldsymbol{F}\boldsymbol{Q}^T, \quad \boldsymbol{g}' = \boldsymbol{Q}\boldsymbol{g}\boldsymbol{Q}^T. \tag{5.149}$$

There are times when we would like to keep the material coordinate system fixed as X_α and use a rotated Cartesian system for the spatial description. Then

$$x'_i = Q_{ij} x_j. \tag{5.150}$$

With this,

$$G'_{\alpha\beta} = \frac{\partial x'_i}{\partial X_\alpha}\frac{\partial x'_i}{\partial X_\beta} = Q_{ij} Q_{ik} \frac{\partial x'_j}{\partial X_\alpha}\frac{\partial x'_k}{\partial X_\beta} = G_{\alpha\beta}, \tag{5.151}$$

where we use $Q_{ij} Q_{ik} = \delta_{jk}$,

$$F'_{\alpha i} = \frac{\partial x'_i}{\partial X_\alpha} = Q_{ij} F_{\alpha j}, \quad \boldsymbol{F}' = \boldsymbol{F}\boldsymbol{Q}^T, \tag{5.152}$$

$$g'_{ij} = \frac{\partial X_\alpha}{\partial x_i}\frac{\partial X_\alpha}{\partial x_j} = Q_{im} g_{mn} Q_{jn}, \quad \boldsymbol{g}' = \boldsymbol{Q}\boldsymbol{g}\boldsymbol{Q}^T. \tag{5.153}$$

We call $\boldsymbol{G}$ a Lagrangian tensor as it is invariant during the rotation of the spatial coordinate system, and we call $\boldsymbol{g}$ an Eulerian tensor as it transforms according to the rules for second-rank tensors during the rotation of the spatial frame. The deformation gradient tensor $\boldsymbol{F}$ has a mixed behavior, as though it has one foot in the Lagrangian system and the other in the Eulerian system. Such tensors are called two-point tensors.

Transformation rules can be derived in a similar way if we keep the spatial frame fixed and rotate the material frame.

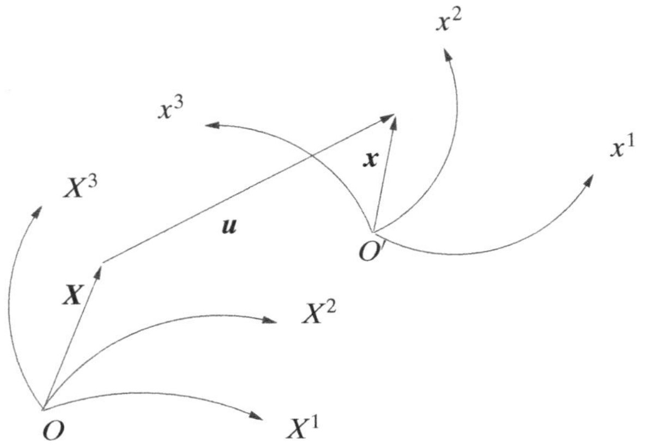

Figure 5.8. Curvilinear coordinates.

5.9 Curvilinear Coordinates

This chapter concludes with a brief introduction to the description of deformation by use of general curvilinear coordinates. As shown in Fig. 5.8, we use two curvilinear coordinates, X^α and x^i, to describe the initial configuration B_0 and the current configuration B. To avoid any confusion with the deformation gradient vectors we saw in the Cartesian description, let $\boldsymbol{C}_\alpha$ and $\boldsymbol{c}_i$ denote the base vectors for the two curvilinear systems. The deformation gradient tensors are

$$F_\alpha{}^i = \frac{\partial x^i}{\partial X^\alpha}, \qquad f_i{}^\alpha = \frac{\partial X^\alpha}{\partial x^i}. \tag{5.154}$$

With

$$C_{\alpha\beta} = \boldsymbol{C}_\alpha.\boldsymbol{C}_\beta, \qquad c_{ij} = \boldsymbol{c}_i.\boldsymbol{c}_j, \tag{5.155}$$

the initial and current distances are given by

$$dS^2 = C_{\alpha\beta} dX^\alpha dX^\beta = C_{\alpha\beta} f_i{}^\alpha f_j{}^\beta dx^i dx^j, \tag{5.156}$$

$$ds^2 = c_{ij} dx^i dx^j = c_{ij} F_\alpha{}^i F_\beta{}^j dX^\alpha dX^\beta. \tag{5.157}$$

The Green's strain tensor $E_{\alpha\beta}$ and the Almansi strain tensor e_{ij} are given by

$$ds^2 - dS^2 = 2E_{\alpha\beta} dX^\alpha dX^\beta = 2e_{ij} dx^i dx^j. \tag{5.158}$$

The displacement vector can be written as

$$\boldsymbol{U} = \boldsymbol{u} = \boldsymbol{C}_\alpha U^\alpha = \boldsymbol{c}_i u^i. \tag{5.159}$$

With the preceding notation, it is easy to recast our earlier Cartesian description into the general curvilinear form. Green and Zerna (1954), Green and Adkins (1960), and Eringen (1962) are excellent references for further study of the general formulations of deformation.

SUGGESTED READING

Eringen, A. C. (1962). *Nonlinear Theory of Continuous Media*, McGraw-Hill.
Green, A. E. and Adkins, J. E. (1960). *Large Elastic Deformations and Nonlinear Continuum Mechanics*, Oxford University Press.
Green, A. E. and Zerna, W. (1954). *Theoretical Elasticity*, Oxford University Press.

EXERCISES

5.1. The deformation of a body is given by the mapping

$$x_1 = X_1 + \epsilon X_2,$$

$$x_2 = -\epsilon X_1 + X_2,$$

$$x_3 = X_3.$$

Obtain the following:

(a) the deformation gradient tensor $F_{\alpha i}$
(b) the deformation tensor $G_{\alpha\beta}$
(c) the Jacobian J
(d) the strain tensor, $E_{\alpha\beta}$
(e) the maximum and minimum stretches and their directions
(f) the right-stretch tensor $\boldsymbol{U}$, the left-stretch tensor $\boldsymbol{V}$, and the rotation tensor $\boldsymbol{R}$
(g) approximate expressions for the preceding quantities by neglecting quadratic terms in ϵ
(h) the shape of a unit circle, $X_1^2 + X_2^2 = 1$, after the deformation using exact and approximate expressions

5.2. Show that an element $d\boldsymbol{X}$, oriented in a principal direction $\boldsymbol{N}^{(\gamma)}$, undergoes a pure rotation and a pure stretch during deformation. Use the polar decomposition to help with this deduction. Also show that any other element would undergo rotation during the stretching operation in $d\boldsymbol{X} \cdot \boldsymbol{V} \cdot \boldsymbol{R}$.

5.3. A cross-section of a thin rubber tube has the flat initial position

$$Y = 0, \quad X = \cos\theta, \quad 0 \leq \theta < 2\pi,$$

where we assume that the thickness of the tube is negligible. If it is deformed into an ellipse,

$$x = a\cos\theta, \quad y = b\sin\theta,$$

where a and b are constants. Obtain an expression for the tangential stretch of the tube wall. Find the locations of the maximum and minimum stretches.

5.4. In an experiment, a specimen undergoing finite plane strain showed stretches $\Lambda = 0.8$ along the X_1 direction and $\Lambda = 0.6$ along the X_2 direction. To deduce the shear strain, a third measurement of the stretch along the direction 45° between the two coordinate directions is made. This has a value of 0.5. Obtain all three components of strain for this case.

5.5. Expand the compatibility equation

$$\nabla \times (\nabla \times \boldsymbol{E})^T = \boldsymbol{O}.$$

Specialize its components to the plane strain case.

5.6. A rubber tube has an annular cross-section with inner radius a and outer radius b. The tube is radially sliced and rejoined in such a way that the old outer surface is now the inner surface. Assume that the midsurface radius and the tube thickness remain constant. Using polar coordinates R, ϕ for the material description and r, θ for the spatial description, obtain deformation relations

$$r = r(R, \phi), \quad \theta = \theta(R, \phi),$$

assuming the cut is made along $\phi = 0$, for this **eversion**. Also, compute the Green's deformation and strain tensors using the R, ϕ coordinates.

5.7. A plane deformation is given by

$$x_1 = X_1 + X_2, \quad x_2 = X_2 - X_1.$$

Obtain the deformation gradient $\boldsymbol{F}$, deformation tensor $\boldsymbol{G}$, and the factors of $\boldsymbol{F}$, $\boldsymbol{U}$, $\boldsymbol{R}$, and $\boldsymbol{V}$. A unit circle in the undeformed body with four points at $\theta = 0, \pi/2, \pi, 3\pi/2$, marked as A, B, C, D, deforms into a curve under this deformation. Obtain the equation for this curve and identify the locations of the four marked points.

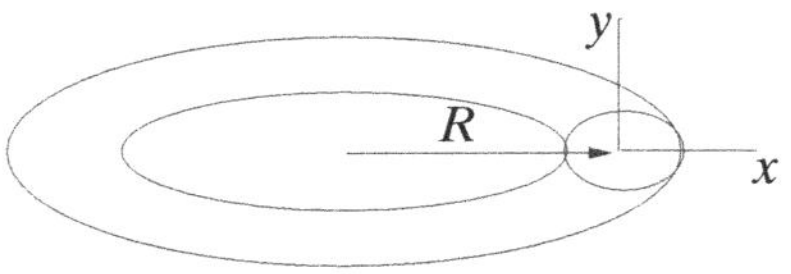

Figure 5.9. Circular ring.

5.8. A circular ring (see Fig. 5.9) has a "neutral" radius R, and its cross-section is spanned by the coordinates x, y. Suppose we apply a uniform, distributed moment around the ring and rotate it by an angle ϕ. Compute the strain of a circular fiber in the ring. Note that, when $\phi = \pi$, the outermost fiber becomes the innermost during the deformation, and the fiber located at the neutral radius does not undergo any strain. Linearize the strains assuming that the angle ϕ is small.

5.9. An infinite thin circular membrane has a hole of radius unity. The hole is closed by deforming it according to

$$x_1 = \left(R + \frac{1}{R}\right)\cos\theta, \quad x_2 = \left(R - \frac{1}{R}\right)\sin\theta,$$

where R and θ are the polar coordinates of a material particle.
Obtain an expression for the stretch of an element tangential to the circle of radius R as a function of θ. Locate the points in the undeformed membrane where these stretches are a maximum and a minimum.

5.10. In a cylindrical coordinate system with coordinates r, θ, and z, and corresponding unit vectors, $\boldsymbol{e}_r$, $\boldsymbol{e}_\theta$, and $\boldsymbol{k}$, the displacement components are u, v, and w. Obtain the strain components in this system and reduce them to the small-strain versions:

$$e_{rr} = \frac{\partial u}{\partial r}, \quad e_{\theta\theta} = \frac{1}{r}\frac{\partial v}{\partial \theta} + \frac{u}{r}, \quad e_{zz} = \frac{\partial w}{\partial z},$$

$$2e_{r\theta} = \frac{1}{r}\frac{\partial u}{\partial \theta} + \frac{\partial v}{\partial r} - \frac{v}{r},$$

$$2e_{\theta z} = \frac{\partial v}{\partial z} + \frac{1}{r}\frac{\partial w}{\partial \theta},$$

$$2e_{zr} = \frac{\partial w}{\partial r} + \frac{\partial u}{\partial z}.$$

Further, if the deformation is axisymmetric, what are the expressions for the small strains?

5.11. In the general curvilinear coordinates, show that the Green's strain and the Almansi strain are given by

$$2E_{\alpha\beta} = D_\alpha U_\beta + D_\beta U_\alpha + D_\alpha U_\gamma D_\beta U^\gamma,$$

$$2e_{ij} = D_i u_j + D_j u_i - D_i u_k D_j u^k,$$

where D_α and D_i are the covariant differential operators in the X^α and x^i coordinate systems, respectively.

5.12. Formulate the eigenvalue problem in the general curvilinear setting for the principal strains when the components of the Green's strain $E_{\alpha\beta}$ or the Almansi strain e_{ij} are given. Note that a unit normal vector $\boldsymbol{N}$ satisfies

$$C_{\alpha\beta} N^\alpha N^\beta = N_\alpha N^\alpha = 1.$$

6 Motion

In the last chapter we considered the initial position $\boldsymbol{X}$ and the final position $\boldsymbol{x}$ of a material particle. The motion of the particle from $\boldsymbol{X}$ to $\boldsymbol{x}$ is a continuous process. This continuity is expressed, with the parameter t representing time, as

$$\boldsymbol{x} = \boldsymbol{x}(X_1, X_2, X_3, t), \quad \boldsymbol{X} = \boldsymbol{X}(x_1, x_2, x_3, t). \tag{6.1}$$

6.1 Material Derivative

In the motion of continuous media the time rates of changes of field quantities associated with the material particles play an important role. Let A, the component of a tensor field, be a function of the material coordinates X_α and time t. The material derivative or substantial derivative of A is defined as

$$\begin{aligned}\frac{DA}{Dt} = \dot{A} &= \lim_{\Delta t \to 0} \frac{1}{\Delta t} \left\{ A\,|_{X,t+\Delta t} - A\,|_{X,t} \right\} \\ &= \frac{\partial A}{\partial t}(X, t),\end{aligned} \tag{6.2}$$

where X stands for X_1, X_2, X_3.

If A is a function of the spatial coordinates x and time t, we have

$$\begin{aligned}\frac{DA}{Dt} = \dot{A} &= \lim_{\Delta t \to 0} \frac{1}{\Delta t} \left\{ A[x_i(X,t),t]\Big|_{X,t+\Delta t} - A[x_i(X,t),t]\Big|_{X,t} \right\} \\ &= \lim_{\Delta t \to 0} \frac{1}{\Delta t} \left\{ A[x_i(X,t) + \frac{\partial x_i}{\partial t}\Big|_{X,t} \Delta t + \cdots, t + \Delta t] - A[x_i(X,t),t] \right\} \\ &= \lim_{\Delta t \to 0} \frac{1}{\Delta t} \left\{ A(x_i, t + \Delta t) + \frac{\partial A}{\partial x_i} \frac{\partial x_i}{\partial t} \Delta t - A(x_i, t) \right\} \\ &= \frac{\partial A}{\partial t}\Big|_{x_i} + \frac{\partial A}{\partial x_i} \frac{\partial x_i}{\partial t}\Big|_{X,t}.\end{aligned}$$

The time derivative of $\boldsymbol{x}$, with the material coordinates fixed, is called the velocity of the particle:

$$\boldsymbol{v} = \frac{\partial \boldsymbol{x}}{\partial t}. \tag{6.3}$$

With this, for any tensor-valued function $\boldsymbol{A}$, we have

$$\frac{D\boldsymbol{A}}{Dt} = \frac{\partial \boldsymbol{A}}{\partial t} + \boldsymbol{v} \cdot \nabla \boldsymbol{A}. \tag{6.4}$$

In the preceding equation, the first term represents the change in $\boldsymbol{A}$ that is due to the change in time at a fixed spatial position. The second term represents the change that is due to the change in $\boldsymbol{x}$ with time. The first term is called the local rate of change, and the second term the convective rate of change. Sums and products of two functions A and B obey

$$\frac{D}{Dt}(A+B) = \frac{DA}{Dt} + \frac{DB}{Dt}, \quad \frac{D}{Dt}(AB) = \frac{DA}{Dt}B + A\frac{DB}{Dt}. \tag{6.5}$$

The acceleration $\boldsymbol{a}$ of a particle is defined as

$$\begin{aligned} \boldsymbol{a} = \dot{\boldsymbol{v}} &= \frac{\partial \boldsymbol{v}}{\partial t} + \boldsymbol{v} \cdot \nabla \boldsymbol{v}, \quad \boldsymbol{v} = \boldsymbol{v}(x, t), \\ &= \frac{\partial \boldsymbol{v}}{\partial t}, \quad \boldsymbol{v} = \boldsymbol{v}(X, t). \end{aligned} \tag{6.6}$$

If we know the Mth material derivative, we can find the $(M+1)$th by the relation

$$\left(\frac{D}{Dt}\right)^{M+1} A = \frac{\partial}{\partial t}\left(\frac{D}{Dt}\right)^{M} A + \boldsymbol{v} \cdot \nabla \left(\frac{D}{Dt}\right)^{M} A. \tag{6.7}$$

6.1.1 Some Terminology

At a spatial point $\boldsymbol{x}$, if $\boldsymbol{v} = 0$, that point is called a **stagnation point**.

If at a point $\boldsymbol{x}$ the velocity $\boldsymbol{v}$ does not explicitly depend on time, that is,

$$\boldsymbol{v} = \boldsymbol{v}(x), \tag{6.8}$$

the motion is called **steady**. However, if we express $\boldsymbol{x}$ in terms of X and t, we find

$$\boldsymbol{v} = \boldsymbol{v}[x(X, t)], \tag{6.9}$$

with t appearing explicitly.

A particle that was at $\boldsymbol{X}$ at time $t = 0$ is called particle X.

Path Lines: A path line is a curve traversed by a particle X as time t varies. It has the parametric equation

$$\boldsymbol{x} = \boldsymbol{x}(X, t); \quad X_\alpha : \text{fixed}, \quad -\infty < t < \infty. \tag{6.10}$$

When a velocity field is given as $v_i = v_i(x, t)$, the path lines can be obtained by solving the differential system

$$\frac{dx_i}{dt} = v_i, \quad x_i|_{t=0} = X_\alpha \delta_{\alpha i}. \tag{6.11}$$

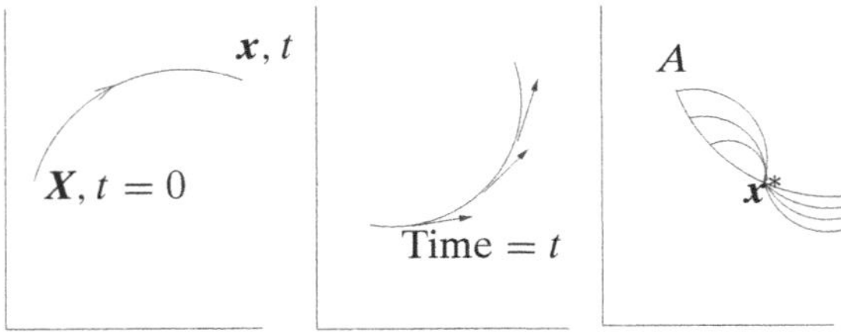

Figure 6.1. Path line, stream line, and streak line.

In Fig. 6.1, the curve on the left shows the path of a particle as time varies.

Stream Lines: At a given time t, the curves tangential to the velocity vectors are called stream lines. That is,

$$d\boldsymbol{x} = k\boldsymbol{v} \tag{6.12}$$

or

$$\frac{dx_1}{v_1} = \frac{dx_2}{v_2} = \frac{dx_3}{v_3}. \tag{6.13}$$

Obviously, at a given time, there is no velocity normal to the stream lines. In Fig. 6.1, the middle diagram shows a stream line with tangential velocity vectors.

Streak Lines: Streak lines are curves representing the locus of material particles at time t that would pass through or would have passed through a given point $\boldsymbol{x}^*$. The particle that would be at $\boldsymbol{x}^*$ at some time τ is identified as

$$X^* = X(x^*, \tau). \tag{6.14}$$

As the parameter τ traverses from the past to the future, we collect all the particles that would ever visit $\boldsymbol{x}^*$. Their location at a particular time t is obtained as

$$\boldsymbol{x} = \boldsymbol{x}(X^*, t) = \boldsymbol{x}[X(x^*, \tau), t]. \tag{6.15}$$

In Fig. 6.1, the curve AB represents the location of the particles at time t that would pass through $\boldsymbol{x}^*$ at some time, along with their paths.

Material curve: We may express parametrically a contiguous curve of material particles as

$$X_\alpha = X_\alpha(s), \tag{6.16}$$

where s is a parameter. This set is called a material curve.

Material surface: With two parameters, s_1 and s_2, we may define a material surface:

$$X_\alpha = X_\alpha(s_1, s_2). \tag{6.17}$$

These material curves and surfaces flow with the motion and occupy spatial curves and surfaces given by

$$x_i = x_i[X(s), t)],$$
$$x_i = x_i[X(s_1, s_2), t]. \tag{6.18}$$

Transient spatial curve: An equation of the form

$$f(x, t) = 0 \tag{6.19}$$

describes a spatial curve that changes with time. If time changes by Δt, the coordinates have to increase by Δx_i to meet

$$\Delta t \frac{\partial f}{\partial t} + \Delta x_i \partial_i f = 0. \tag{6.20}$$

Dividing by Δt and taking the limit, we find

$$\frac{\partial f}{\partial t} + \frac{\partial x_i}{\partial t} \partial_i f = 0. \tag{6.21}$$

At any instant this gives the rate of change of the coordinates of a point on this curve. If it is a material curve, this rate of change will be the velocity of the material particle, that is,

$$\frac{\partial x_i}{\partial t} = v_i. \tag{6.22}$$

From this we get the Lagrange criterion for a spatial curve to be a material curve:

$$\frac{Df}{Dt} = \frac{\partial f}{\partial t} + \boldsymbol{v} \cdot \nabla f = 0. \tag{6.23}$$

6.1.2 Example: Path Line, Stream Line, and Streak Line

Consider the plane motion given by the equations

$$x_1 = X_1 \cosh t + X_2 \sinh t, \quad x_2 = X_1 \sinh t + X_2 \cosh t. \tag{6.24}$$

1. Obtain the path line of a particle that was at the point (0,1) at time $t = 0$. Using $t = 0$ and $x_1 = 0$, $x_2 = 1$ in Eqs. (6.24), we find $X_1 = 0$, $X_2 = 1$. The equation for the path line (see Fig. 6.2) has the parametric form

   $$x_1 = \sinh t, \quad x_2 = \cosh t,$$

   which represents the hyperbola

   $$x_2^2 - x_1^2 = 1. \tag{6.25}$$

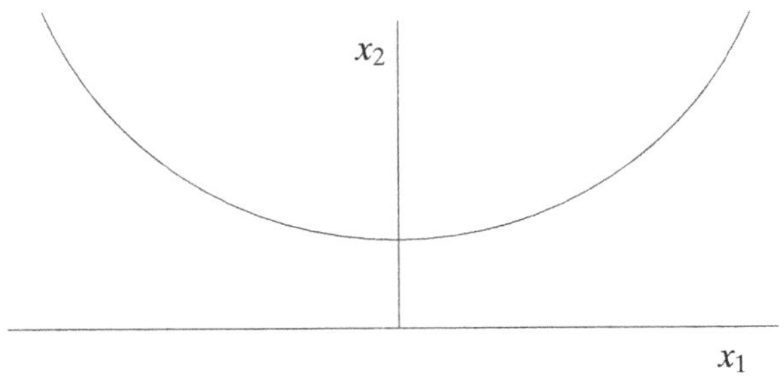

Figure 6.2. Path line.

2. Obtain the stream line through (0,1) at time $t = 0$.
   From Eqs. (6.24), the velocity components are

$$v_1 = X_1 \sinh t + X_2 \cosh t, \quad v_2 = X_1 \cosh t + X_2 \sinh t,$$

   Eliminating X_1 and X_2, by using Eqs. (6.24) we obtain

$$v_1 = x_2, \quad v_2 = x_1.$$

   As stream lines are tangential to the velocity vector,

$$\frac{dx_1}{v_1} = \frac{dx_2}{v_2} \quad \text{or} \quad x_1 dx_1 - x_2 dx_2 = 0.$$

   Integrating this, we find $x_2^2 - x_1^2 = C$. The condition $x_1 = 0, x_2 = 1$ yields the hyperbola

$$x_2^2 - x_1^2 = 1. \tag{6.26}$$

3. Obtain the streak line through (0,1) when $t = 1$.
   To identify all the particles that would visit the spatial location (0,1) in time τ, we invert Eqs. (6.24) and substitute $x_1 = 0, x_2 = 1$ to get

$$X_1 = -\sinh \tau, \quad X_2 = \cosh \tau.$$

   Their location at time t is given by

$$x_1 = -\cosh t \sinh \tau + \sinh t \cosh \tau = \sinh(t - \tau),$$

$$x_2 = \cosh t \cosh \tau - \sinh t \sinh \tau = \cosh(t - \tau).$$

   Using $t = 1$ and eliminating $(1 - \tau)$, we have, again,

$$x_2^2 - x_1^2 = 1. \tag{6.27}$$

Of course, for a steady flow given by Eqs. (6.24), all three lines are identical. This example illustrates the procedure involved in finding these lines from the given motion.

6.2 Length, Volume, and Area Elements

In this section we are interested in finding the material derivatives of elements of arc length, area, and volume.

With $x_i = x_i(X)$, we begin by computing

$$\frac{D}{Dt}(\partial_\alpha x_i) = \partial_\alpha v_i = \partial_j v_i \partial_\alpha x_j. \tag{6.28}$$

By removing ∂_α, we find

$$\frac{D}{Dt}(dx_i) = \partial_j v_i dx_j. \tag{6.29}$$

From the identity

$$\partial_i X_\alpha \partial_\beta x_i = \delta_{\alpha\beta},$$

we get

$$
\begin{aligned}
\frac{D}{Dt}(\partial_i X_\alpha \partial_\beta x_i) &= 0, \\
\partial_\beta x_i \frac{D}{Dt}(\partial_i X_\alpha) &= -\partial_i X_\alpha \frac{D}{Dt}(\partial_\beta x_i), \\
\partial_\beta x_i \frac{D}{Dt}(\partial_i X_\alpha) &= -\partial_j X_\alpha \partial_i v_j \partial_\beta x_i, \\
\frac{D}{Dt}(\partial_i X_\alpha) &= -\partial_i v_j \partial_j X_\alpha.
\end{aligned} \tag{6.30}
$$

6.2.1 Length

We have

$$\frac{D}{Dt}(ds)^2 = \frac{D}{Dt}(dx_i dx_i) = 2\partial_i v_j dx_i dx_j. \tag{6.31}$$

The velocity gradient

$$l_{ij} = \partial_i v_j \tag{6.32}$$

can be resolved into a symmetric and a skew-symmetric component as

$$\partial_i v_j = l_{ij} = d_{ij} + w_{ij}, \tag{6.33}$$

where

$$d_{ij} = \frac{1}{2}(l_{ij} + l_{ji}), \quad w_{ij} = \frac{1}{2}(l_{ij} - l_{ji}). \tag{6.34}$$

The tensors $\boldsymbol{d}$ and $\boldsymbol{w}$ are known as the deformation rate and the spin tensors, respectively. As $dx_i dx_j$ is symmetric, we get

$$\frac{D}{Dt}(ds)^2 = 2d_{ij} dx_i dx_j. \tag{6.35}$$

6.2.2 Volume

From

$$dv = J dV, \tag{6.36}$$

we get

$$\frac{D}{Dt}(dv) = \frac{DJ}{Dt} dV, \tag{6.37}$$

where

$$
\begin{aligned}
\frac{DJ}{Dt} &= \frac{\partial J}{\partial(\partial_\alpha x_i)} \frac{D}{Dt}(\partial_\alpha x_j) \\
&= \frac{\partial J}{\partial(\partial_\alpha x_i)} \partial_\alpha x_j \partial_j v_i.
\end{aligned} \tag{6.38}
$$

Observing that the partial derivative of J is the cofactor in the determinant expansion for J, we obtain

$$\frac{\partial J}{\partial(\partial_\alpha x_i)} = J\,\partial_i X_\alpha. \tag{6.39}$$

Then

$$\frac{DJ}{Dt} = J\,\partial_i v_i, \quad \frac{D}{Dt}(dv) = \partial_i v_i dv. \tag{6.40}$$

6.2.3 Area

The material derivative of an area element, da_i, can be obtained as

$$\begin{aligned}\frac{D}{Dt}(da_i) &= \frac{D}{Dt}(J\,dA_\alpha \partial_i X_\alpha)\\ &= \left[\frac{DJ}{Dt}\partial_i X_\alpha + J\frac{D}{Dt}(\partial_i X_\alpha)\right] dA_\alpha\\ &= \partial_j v_j da_i - \partial_i v_j da_j.\end{aligned} \tag{6.41}$$

6.3 Material Derivatives of Integrals

We come across different types of integral quantities in continuum mechanics, and it is important to know how to calculate the rate of change of these integrals.

6.3.1 Line Integrals

Let C_0 be a material curve at $t = 0$ and C be its deformed shape at time t. The material derivative of the line integral of ϕ, which is a component of a tensor-valued function, can be written as

$$\begin{aligned}\frac{D}{Dt}\int_C \phi(x)dx_i &= \frac{D}{Dt}\int_{C_0} \phi(x)\partial_\alpha x_i dX_\alpha\\ &= \int_{C_0} \frac{D}{Dt}(\phi\partial_\alpha x_i)dX_\alpha\\ &= \int_{C_O}\left[\partial_\alpha x_i \frac{D\phi}{Dt} + \phi\partial_j v_i \partial_\alpha x_j\right] dX_\alpha\\ &= \int_C \left[\frac{D\phi}{Dt}dx_i + \phi\partial_j v_i dx_j\right].\end{aligned} \tag{6.42}$$

If C is a fixed curve in space,

$$\frac{d}{dt}\int_C \phi dx_i = \int_C \frac{\partial\phi}{\partial t}dx_i. \tag{6.43}$$

6.3.2 Area Integrals

For a material surface S, using the transformation to the undeformed system, we get the material derivative of the area integral of ϕ:

$$\frac{D}{Dt}\int_S \phi(x)da_i = \int_S \left[\frac{D\phi}{Dt}da_i + \phi(\partial_j v_j da_i - \partial_i v_j da_j)\right]. \tag{6.44}$$

If the surface S is fixed in space,

$$\frac{d}{dt}\int_S \phi da_i = \int_S \frac{\partial \phi}{\partial t}da_i. \tag{6.45}$$

6.3.3 Volume Integrals

If V is a material volume, the material derivative of the volume integral is obtained as

$$\begin{aligned}\frac{D}{Dt}\int_V \phi(x)dv &= \int_V \left[\frac{D\phi}{Dt} + \phi\partial_i v_i\right]dv \\ &= \int_V \left[\frac{\partial \phi}{\partial t} + \partial_i(\phi v_i)\right]dv.\end{aligned} \tag{6.46}$$

For a spatial volume,

$$\frac{d}{dt}\int_V \phi dv = \int_V \frac{\partial \phi}{\partial t}dv. \tag{6.47}$$

6.4 Deformation Rate, Spin, and Vorticity

We have introduced the deformation rate tensor $\boldsymbol{d}$ and the spin tensor $\boldsymbol{w}$ as

$$\boldsymbol{d} = \frac{1}{2}[\nabla \boldsymbol{v} + (\nabla \boldsymbol{v})^T], \quad \boldsymbol{w} = \frac{1}{2}[\nabla \boldsymbol{v} - (\nabla \boldsymbol{v})^T]. \tag{6.48}$$

The skew-symmetric tensor $\boldsymbol{w}$ has three nonzero components:

$$w_{12} = -w_{21}, \quad w_{23} = -w_{32}, \quad w_{31} = -w_{13}. \tag{6.49}$$

We can associate a vector $\boldsymbol{\omega}$, called a dual vector, with the skew-symmetric tensor, with its components defined as

$$\omega_i = e_{ijk}w_{jk} = e_{ijk}\partial_j v_k \tag{6.50}$$

or

$$\boldsymbol{\omega} = \nabla \times \boldsymbol{v}. \tag{6.51}$$

The vector $\boldsymbol{\omega}$ is called the **vorticity** vector. In particular,

$$\omega_1 = 2w_{23}, \quad \omega_2 = 2w_{31}, \quad \omega_3 = 2w_{12}. \tag{6.52}$$

For a physical interpretation of $\boldsymbol{d}$, we begin by defining the **stretching** of an element, $d\boldsymbol{x}$, oriented in the direction $\boldsymbol{n}$ in the current configuration as

$$\begin{aligned} d_n &= \frac{1}{ds}\frac{D}{Dt}(ds) = \frac{1}{2}\frac{1}{(ds)^2}\frac{D}{Dt}(ds)^2, \\ &= d_{ij}\frac{dx_i}{ds}\frac{dx_j}{ds} = d_{ij}n_i n_j. \end{aligned} \tag{6.53}$$

This shows that the diagonal elements d_{11}, d_{22}, and d_{33} are the rates at which elements pointing in the 1-, 2-, and 3-directions stretch.

To see the significance of the off-diagonal elements of $\boldsymbol{d}$, consider two material elements $d\boldsymbol{x}$ and $d\boldsymbol{x}^*$ oriented in the directions $\boldsymbol{n}$ and $\boldsymbol{n}^*$, respectively (see Fig. 6.3). If θ is the angle between them,

$$\cos\theta = n_i n_i^* = \frac{dx_i}{ds}\frac{dx_i^*}{ds^*}, \tag{6.54}$$

$$\begin{aligned} \frac{D}{Dt}(\cos\theta) &= \frac{D}{Dt}\left(\frac{dx_i}{ds}\frac{dx_i^*}{ds}\right) \\ &= \frac{dx_i}{ds ds^*}\frac{D}{Dt}(dx_i^*) + \frac{dx_i^*}{ds ds^*}\frac{D}{Dt}(dx_i) \\ &\quad - \frac{dx_i dx_i^*}{(ds)^2 ds^*}\frac{D}{Dt}(ds) - \frac{dx_i dx_i^*}{ds(ds^*)^2}\frac{D}{Dt}(ds^*) \\ &= 2d_{ij}\frac{dx_i}{ds}\frac{dx_j^*}{ds^*} - [d_{n^*} + d_n]\frac{dx_i}{ds}\frac{dx_i^*}{ds^*}, \\ -\sin\theta\frac{D\theta}{Dt} &= 2d_{ij}n_i n_j^* - [d_n + d_{n^*}]\cos\theta. \end{aligned} \tag{6.55}$$

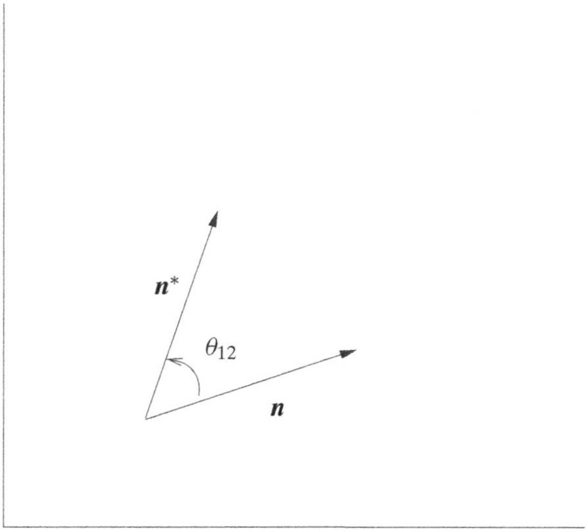

Figure 6.3. Shearing deformation of the angle between two material elements.

In the special case, $\boldsymbol{n} = \boldsymbol{e}_1$ and $\boldsymbol{n}^*=\boldsymbol{e}_2$, we have $\theta = \pi/2$ and

$$-\frac{D\theta}{Dt} = 2d_{12}. \tag{6.56}$$

Thus $2d_{12}$ represents the rate of closing of the 90° angle between $\boldsymbol{e}_1$ and $\boldsymbol{e}_2$. We refer to d_{12} as the mathematical shearing strain. To interpret the spin components, we consider a fixed unit vector $\boldsymbol{\nu}$ in space and a unit vector $\boldsymbol{n}$ formed by material particles. Let ϕ denote the angle between them. Then,

$$\cos\phi = \nu_i n_i = \nu_i \frac{dx_i}{ds} \tag{6.57}$$

$$\begin{aligned} \frac{D}{Dt}\cos\phi &= \nu_i \frac{D}{Dt}\left(\frac{dx_i}{ds}\right), \\ -\sin\phi \frac{D\phi}{Dt} &= \nu_i \left[\partial_j v_i \frac{dx_j}{ds} - \frac{dx_i}{(ds)^2}\frac{D}{Dt}(ds)\right] \\ &= \nu_i [n_j \partial_j v_i - n_i d_n]. \end{aligned} \tag{6.58}$$

There are two special cases:

1. Assuming

$$\boldsymbol{\nu} = \boldsymbol{e}_1, \quad \boldsymbol{n} = \boldsymbol{e}_2, \tag{6.59}$$

   we get

$$\frac{D\phi_{21}}{Dt} = -\partial_2 v_1. \tag{6.60}$$

2. Assuming

$$\boldsymbol{\nu} = \boldsymbol{e}_2, \quad \boldsymbol{n} = \boldsymbol{e}_1, \tag{6.61}$$

   we get

$$\frac{D\phi_{12}}{Dt} = -\partial_1 v_2. \tag{6.62}$$

From the expression for ϕ_{21} we observe that, if v_1 increases with x_2, the vertical material element gets rotated clockwise. The velocity component v_2 is responsible for rotating a horizontal element, and if it increases with x_1 an anticlockwise rotation is imparted. Here, we use the notation ϕ_{21} to indicate the angle from fixed $\boldsymbol{e}_1$ to rotating $\boldsymbol{e}_2$ and ϕ_{12} when the two unit vectors are interchanged (see Fig. 6.4).

With these, we can express the deformation rate and spin components in the form

$$2d_{12} = -(\dot{\phi}_{12} + \dot{\phi}_{21}), \quad 2w_{12} = \dot{\phi}_{21} - \dot{\phi}_{12}. \tag{6.63}$$

Figure 6.5 shows the velocity distribution of a flow:

$$v_1 = -\frac{1}{2}\Omega x_2, \quad v_2 = \frac{1}{2}\Omega x_1. \tag{6.64}$$

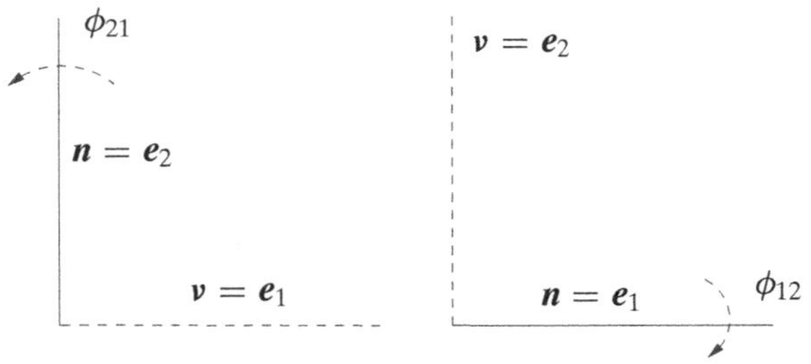

Figure 6.4. Rotation of material elements relative to fixed directions.

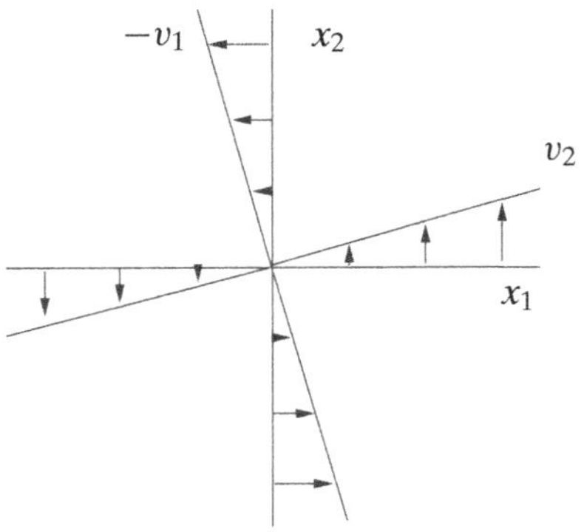

Figure 6.5. Instantaneous rotation of a flow.

From this,

$$\partial_1 v_2 = \frac{1}{2}\Omega, \quad \partial_2 v_1 = -\frac{1}{2}\Omega, \tag{6.65}$$

$$d_{12} = 0, \quad w_{12} = \frac{1}{2}\Omega. \tag{6.66}$$

The corresponding dual vector (vorticity) is obtained as

$$\boldsymbol{\Omega} = \Omega \boldsymbol{e}_3. \tag{6.67}$$

6.5 Strain Rate

Let us begin by noting that

$$\frac{D}{Dt}(ds)^2 = 2d_{ij}dx_i dx_j, \tag{6.68}$$

$$\frac{D}{Dt}[(ds)^2 - (dS)^2] = 2\dot{E}_{\alpha\beta}dX_\alpha dX_\beta, \quad \frac{D}{Dt}(dS)^2 = 0. \tag{6.69}$$

From these,

$$\dot{E}_{\alpha\beta} = \partial_\alpha x_i \partial_\beta x_j d_{ij}. \tag{6.70}$$

The material derivative of the Almansi strain follows from

$$\frac{D}{Dt}[(ds)^2 - (dS)^2] = 2\frac{D}{Dt}(e_{ij}dx_i dx_j), \tag{6.71}$$

as

$$\dot{e}_{ij} = d_{ij} - \partial_i v_k e_{kj} - e_{ik}\partial_j v_k. \tag{6.72}$$

We also have

$$\dot{G}_{\alpha\beta} = 2\dot{E}_{\alpha\beta}, \quad \dot{g}_{ij} = -2\dot{e}_{ij}. \tag{6.73}$$

6.6 Rotation Rate of Principal Axis

In Chapter 5 we saw that a principal direction of the spatial ellipsoid $\boldsymbol{N}$ gets mapped into a principal direction of the material ellipsoid $\boldsymbol{n}$. Using the polar decomposition, we obtain

$$\boldsymbol{n} = \boldsymbol{R}^T\boldsymbol{N}. \tag{6.74}$$

The rotation rate of this unit vector is given by

$$\dot{\boldsymbol{n}} = \dot{\boldsymbol{R}}^T\boldsymbol{N}. \tag{6.75}$$

Eliminating $\boldsymbol{N}$, we have

$$\dot{\boldsymbol{n}} = \dot{\boldsymbol{R}}^T\boldsymbol{R}\boldsymbol{n} = \boldsymbol{W}_T\boldsymbol{n}, \tag{6.76}$$

where $\boldsymbol{W}_T$ is called the **twirl** tensor.

SUGGESTED READING

Batchelor, G. K. (1967). *An Introduction to Fluid Dynamics*, Cambridge University Press.
Bird, R. B., Stewart, W. E., and Lightfoot, E. N. (1960). *Transport Phenomena*, Wiley.

EXERCISES

6.1. A steady flow is defined by the relation $\boldsymbol{v} = \boldsymbol{v}(x)$. Show that the path lines, stream lines, and streak lines through a specified point coincide for this case.

6.2. Circulation of a fluid around a closed material curve C is defined as

$$\Gamma = \oint_C \boldsymbol{v} \cdot d\boldsymbol{x}.$$

(a) Transform this integral into an area integral using Stokes theorem.
(b) A motion is circulation preserving if $\dot{\Gamma} = \boldsymbol{0}$. Show that, in that case,

$$\frac{\partial \boldsymbol{\omega}}{\partial t} + \nabla \times (\boldsymbol{\omega} \times \boldsymbol{v}) = \boldsymbol{0}.$$

(c) Deduce from this

$$\nabla \times \dot{\boldsymbol{v}} = \boldsymbol{0}.$$

(d) Show that the expression for acceleration,

$$\dot{\boldsymbol{v}} = \nabla \phi,$$

satisfies the preceding equation.

6.3. A flow is irrotational if $\boldsymbol{w} = \boldsymbol{0}$. Show that this condition is satisfied if $\boldsymbol{v}$ is written as

$$\boldsymbol{v} = \nabla \phi,$$

where ϕ is called the velocity potential.

6.4. A flow is incompressible if $DJ/Dt = 0$. Show that this condition is satisfied if $\boldsymbol{v}$ is written as

$$\boldsymbol{v} = \nabla \times \boldsymbol{\psi}.$$

6.5. For a plane ($v_3 = 0$), incompressible flow, we may set $\boldsymbol{\psi} = \psi \boldsymbol{e}_3$, where ψ is called the stream function. Show that, for incompressible, irrotational flow,

$$\partial_1 \phi = \partial_2 \psi, \quad \partial_2 \phi = -\partial_1 \psi,$$

$$\nabla^2 \phi = 0, \quad \nabla^2 \psi = 0.$$

6.6. For a two-dimensional (2D) incompressible flow ($v_3 = 0$) around a unit circle with center at $r = 0$, the velocity components are

$$v_1 = U\left(1 - \frac{\cos 2\theta}{r^2}\right), \quad v_2 = -U\frac{\sin 2\theta}{r^2},$$

where U is a constant representing the uniform velocity of the fluid in the far field and r and θ are polar coordinates. Show that the component of velocity normal to the circular boundary is zero. Compute the rate of volume change and vorticity. Obtain the location of the maximum rate of change of the kinetic energy density, DK/Dt, for this flow. The kinetic energy density is defined as $K = \rho v^2/2$.

6.7. In a plane flow, fluid flows out of a source radially, with the radial velocity

$$v_r = \frac{q}{2\pi r},$$

where q is the source strength. An expanding space curve imposed on this flow has the equation

$$r - f(t) = 0.$$

Obtain $f(t)$ so that this space curve is a material curve.

6.8. The equation for a material curve in a flow is given by

$$x - ty - t^2 = 0,$$

where t represents time. The velocity component v_y has a constant value, 2. Obtain the component v_x, the deformation rate tensor $\boldsymbol{d}$, and the spin tensor $\boldsymbol{w}$.

6.9. The motion of a fluid is described by

$$v_1 = x_2(V_0 - \Omega t), \quad v_2 = x_1(-V_0 + \Omega t), \quad v_3 = 0.$$

Compute the deformation rate, spin, and vorticity for this flow. What is the equation for the stream line through (1,1,0) at $t = 1$? Is this a steady flow?

6.10. In a plane flow the velocity components are

$$v_1 = U\frac{x_2}{a}, \qquad v_2 = -U\frac{x_1}{a},$$

where U and a are constants. Obtain the stream line through the point (a, a). At time $t = 0$, a material line in this flow has the equation $x_1 - x_2 = 0$. Obtain the equation for this line at $t = 1$.

7 Fundamental Laws of Mechanics

The fundamental laws of mechanics dealing with mass, momentum, energy, and entropy are stated and briefly discussed in this chapter. These concepts have long histories, and it is not necessary or appropriate in our study of continuum mechanics to elaborate on these topics. Our discussion is based on the elementary ideas we have gathered from our earlier studies of statics, dynamics, and thermodynamics.

7.1 Mass

With each body we associate a measure called mass, which is nonnegative and additive. It is assumed to be invariant under the class of motions encountered in engineering. With a lower limit on the length scales mentioned in Chapter 1, we treat mass as continuous and define the mass density ρ as

$$\rho = \lim_{\Delta V \to 0} \frac{\Delta m}{\Delta V}. \tag{7.1}$$

With this, the mass of a body occupying a volume V can be expressed as

$$m = \int_V \rho dv, \tag{7.2}$$

and integrals

$$\int_m f dm = \int_V f \rho dv. \tag{7.3}$$

The following mass integrals play important roles in mechanics:

1. Total mass:

$$m = \int dm. \tag{7.4}$$

2. Linear momentum:

$$\mathbf{L} = \int_m \mathbf{v} dm. \tag{7.5}$$

3. Moment of momentum or angular momentum:

$$\boldsymbol{H} = \int_m \boldsymbol{x} \times \boldsymbol{v} dm. \tag{7.6}$$

Although this relation is based on the assumption that the moment is taken about the origin, we are free to select any convenient point.

4. Kinetic energy:

$$K = \frac{1}{2} \int_m \boldsymbol{v} \cdot \boldsymbol{v} dm. \tag{7.7}$$

5. Internal energy:

$$E = \int_m \varepsilon dm, \tag{7.8}$$

where ε is the internal energy density.

6. Entropy:

$$S = \int_m s dm, \tag{7.9}$$

where s is the entropy density.

We will modify some of these integrals to include microstructural details with additional degrees of freedom such as microrotations, on top of the three displacement components.

7.2 Conservation and Balance Laws

7.2.1 Conservation of Mass

The law of conservation of mass states that total mass of a body remains invariant under any motion:

$$m = \int_{V_0} \rho_0 dV_0 = \int_V \rho dV. \tag{7.10}$$

7.2.2 Balance of Linear Momentum

The time rate of change of linear momentum is equal to the applied force $\boldsymbol{F}$ acting on the body:

$$\frac{D\boldsymbol{L}}{Dt} = \boldsymbol{F} \Leftrightarrow \frac{D}{Dt} \int_V \boldsymbol{v} \rho dV = \boldsymbol{F}. \tag{7.11}$$

7.2.3 Balance of Angular Momentum

The time rate of change of angular momentum about a fixed point is equal to the applied moment about the same fixed point $\boldsymbol{M}$:

$$\frac{D\boldsymbol{H}}{Dt} = \boldsymbol{M} \Leftrightarrow \frac{D}{Dt}\int_V [\boldsymbol{x} \times \boldsymbol{v} + \boldsymbol{J} \cdot \boldsymbol{\mu}]\rho dV = \boldsymbol{M}. \tag{7.12}$$

Here, we modify the classical moment of momentum, $\rho \boldsymbol{x} \times \boldsymbol{v}$, by adding a microstructural spin angular momentum, $\rho \boldsymbol{J} \cdot \boldsymbol{\mu}$, with $\rho \boldsymbol{J}$ representing inertia and $\boldsymbol{\mu}$ the angular velocity. We will discuss the significance of this complication in Chapter 9. The two balance laws previously stated (without the complication) in the global form are called the Euler equations of motion.

7.2.4 Balance of Energy

The time rate of change of the sum of the kinetic and the internal energies is equal to the rate of flow of energies into the body. Of all possible energies flowing into the body, we concentrate mainly on the mechanical power input and the rate of heat energy entering the body:

$$\frac{D}{Dt}(K + E) = P + H. \tag{7.13}$$

where P is the mechanical power input and H is the rate of flow of heat energy. This statement is also known as the first principle of thermostatics.

7.2.5 Entropy Production

In any irreversible process taking place in a closed system, the rate of entropy produced has to be positive. The total entropy change can be written as the sum of the entropy change that is due to reversible processes S_E and irreversible processes S_I:

$$S = S_E + S_I. \tag{7.14}$$

The entropy production inequality or the second law of thermostatics states that

$$\frac{DS_I}{Dt} \geq 0, \tag{7.15}$$

with the equality applicable if there is no irreversible process involved. We will discuss this principle in more detail in Chapter 9.

7.3 Axiom of Material Frame Indifference

We discussed the motion of a continuum with respect to a coordinate system fixed in space. In classical mechanics we consider the Newtonian ideal of absolute space and absolute time and an observer with a measuring tape and a clock. For everyday engineering applications (including trips to the Moon and beyond) this approach has worked out satisfactorily. When the physical properties of materials are to be determined, we should know how the motion of the observer will affect the measurements. Although the values of the measured quantities may depend on the motion of the observer, their interrelations must be independent of the observer. For example, the velocity and acceleration of a particle would appear differently to different observers moving with their own velocities and accelerations. However, the distance ds the particle covers in time dt must be independent of the observers. Quantities that are invariant under observer transformations are called **frame indifferent**. Other terms for this concept include **objectivity** and **observer independence**. An **observer transformation** is defined as

$$x'_i = Q_{ij}(t)x_j + b_i(t), \tag{7.16}$$

where x and t represent an absolute Cartesian system of coordinates and time, as in classical mechanics. The observer's coordinate system x' is translating and rotating as a function of time. The rotation matrix $\boldsymbol{Q}(t)$ is orthogonal.

The distance relation is

$$(ds)^2 = dx'_i dx'_i = Q_{im}Q_{in}dx_m dx_n = \delta_{mn}dx_m dx_n = dx_m dx_m. \tag{7.17}$$

Thus the absolute distance and the distance measured by the observer are identical.

It is useful to note that

$$\boldsymbol{Q}^T\boldsymbol{Q} = \boldsymbol{I}, \tag{7.18}$$

and, by differentiating, we obtain

$$\boldsymbol{Q}^T\dot{\boldsymbol{Q}} = -\dot{\boldsymbol{Q}}^T\boldsymbol{Q}. \tag{7.19}$$

Including a constant shift in time $t' = t - a$, we define two motions, $x'_i(X, t')$ and $x_j(X, t)$, as objectively equivalent if

$$x'_i(X, t') = Q_{ij}(t)x_j(X, t) + b_i(t). \tag{7.20}$$

The relation between velocities,

$$\frac{D}{Dt'}x'_i(X, t') = Q_{ij}\frac{D}{Dt}x_j(X, t) + \dot{Q}_{ij}x_j(X, t) + \dot{b}_i(t),$$
$$v'_i(X, t') = Q_{ij}v_j(X, t) + \dot{Q}_{ij}x_j(X, t) + \dot{b}_i, \tag{7.21}$$

shows that velocity components are not frame indifferent.

Differentiating again, we have the acceleration

$$a'_i = Q_{ij}a_j + 2\dot{Q}_{ij}v_j + \ddot{Q}_{ij}x_j + \ddot{b}_i, \tag{7.22}$$

which is, again, not frame indifferent.

An objective tensor would follow the transformation law for tensors under objective coordinate transformations. That is, A_{ij} is objective if

$$A'_{ij}(X, t') = Q_{im}(t)Q_{jn}(t)A_{mn}(X, t). \tag{7.23}$$

Using the relations

$$dx'_i = Q_{ij}dx_j, \quad dx_j = Q_{ij}dx'_i, \tag{7.24}$$

we can write the velocity gradient, which is not objective, as

$$\begin{aligned}\partial_k' v_i' &= Q_{ij}\partial_l v_j \frac{\partial x_l}{\partial x_k'} + \dot{Q}_{ij}\frac{\partial x_j}{\partial x_k'}\\ &= Q_{ij}Q_{kl}\partial_l v_j + \dot{Q}_{ij}Q_{kj}. \end{aligned} \tag{7.25}$$

But the deformation rate becomes

$$d_{ik}' = \frac{1}{2}(\partial_k' v_i' + \partial_i' v_k') = Q_{ij}Q_{kl}d_{jl} + \frac{1}{2}(\dot{Q}_{ij}Q_{kj} + Q_{ij}\dot{Q}_{kj}), \tag{7.26}$$

where the second term is zero. Thus $\boldsymbol{d}$ is objective as

$$d_{ij}' = Q_{im}Q_{jn}d_{mn}. \tag{7.27}$$

If we assemble the spin tensor, we find

$$w_{ij}' = Q_{im}Q_{jn}w_{mn} + \frac{1}{2}(Q_{ik}\dot{Q}_{jk} - \dot{Q}_{ik}Q_{jk}), \tag{7.28}$$

which is not objective.

In matrix form

$$\begin{aligned}\boldsymbol{w}' &= \boldsymbol{Q}\boldsymbol{w}\boldsymbol{Q}^T + \frac{1}{2}(\boldsymbol{Q}\dot{\boldsymbol{Q}}^T - \dot{\boldsymbol{Q}}\boldsymbol{Q}^T)\\ &= \boldsymbol{Q}\boldsymbol{w}\boldsymbol{Q}^T + \boldsymbol{Q}\dot{\boldsymbol{Q}}^T. \end{aligned} \tag{7.29}$$

From this we obtain an expression for $\dot{\boldsymbol{Q}}$ in terms of the spins in the two systems:

$$\dot{\boldsymbol{Q}} = \boldsymbol{Q}\boldsymbol{w} - \boldsymbol{w}'\boldsymbol{Q}. \tag{7.30}$$

7.4 Objective Measures of Rotation

In applications involving viscoelastic polymers, objective measures of rotation have been found to be useful. One such measure is the objective part of the spin tensor (VanArsdale, 2003; Zhou and Tamma, 2003). Consider the velocity gradient in the form

$$\begin{aligned}\nabla \boldsymbol{v} &= \frac{\partial v_j}{\partial x_i}\boldsymbol{e}_i\boldsymbol{e}_j\\ &= \frac{\partial X_\alpha}{\partial x_i}\frac{D}{Dt}\frac{\partial x_j}{\partial X_\alpha}\boldsymbol{e}_i\boldsymbol{e}_j\\ &= \boldsymbol{f}\dot{\boldsymbol{F}} \qquad (7.31)\\ &= \boldsymbol{R}^T\boldsymbol{V}^{-1}[\dot{\boldsymbol{V}}\boldsymbol{R} + \boldsymbol{V}\dot{\boldsymbol{R}}]\\ &= \boldsymbol{R}^T\dot{\boldsymbol{R}} + \boldsymbol{R}^T\boldsymbol{V}^{-1}\dot{\boldsymbol{V}}\boldsymbol{R}. \qquad (7.32)\end{aligned}$$

Using the transpose of this expression, we find

$$\boldsymbol{d} = \frac{1}{2}\boldsymbol{R}^T[\boldsymbol{V}^{-1}\dot{\boldsymbol{V}} + \dot{\boldsymbol{V}}\boldsymbol{V}^{-1}]\boldsymbol{R}, \tag{7.33}$$

$$\boldsymbol{w} = \boldsymbol{w}_r + \boldsymbol{w}_d, \quad \boldsymbol{w}_r \equiv \boldsymbol{R}^T\dot{\boldsymbol{R}}, \quad \boldsymbol{w}_d \equiv \frac{1}{2}\boldsymbol{R}^T[\boldsymbol{V}^{-1}\dot{\boldsymbol{V}} - \dot{\boldsymbol{V}}\boldsymbol{V}^{-1}]\boldsymbol{R}, \tag{7.34}$$

where $\boldsymbol{w}_r$ represents the rigid body spin and $\boldsymbol{w}_d$ the deformational contribution to the overall spin; the latter part is a frame-independent quantity that is suitable for use in describing material response to stresses.

Wedgewood and Geurts (1995) and Wedgewood (1999) introduced a rigid body rotation rate in a rotating–translating observer frame in such a way that the average components of the quantity $\boldsymbol{v} \times \dot{\boldsymbol{v}}$ is set to zero. They assume the velocity distribution in a small neighborhood in the form

$$\bar{\boldsymbol{v}} = \boldsymbol{r} \cdot \nabla \boldsymbol{v}. \tag{7.35}$$

This approach leads to multiple values for the rotation rate in certain cases, and the boundary conditions have to be used to select the physically feasible solution.

Later, when we discuss the constitutive relations, we restrict the class of variables used to be objective tensors, and these relations must satisfy the fundamental laws of mechanics.

7.5 Integrity Basis

In continuum mechanics we encounter scalar-, vector-, and tensor-valued functions with scalars, vectors, or tensors as arguments. The symmetry properties of the functions under coordinate rotations dictate the allowable forms for the arguments. For a scalar function with scalar variables, we have the function and its variables invariant under a rotation of coordinates. When a scalar function has vectors as arguments, the function can depend on only the invariants obtained from the vectors under arbitrary rotations. These invariants are the norms of the vectors, their inner products taken two at a time and their triple products taken three at a time. This invariant group is called the irreducible integrity basis of the scalar function. This result is attributed to Cauchy.

When a scalar function ψ depends on a second-rank tensor $\boldsymbol{A}$, the arguments take the form of the invariants of $\boldsymbol{A}$, namely, I_{A1}, I_{A2}, and I_{A3}. Because the first invariant of $\boldsymbol{A}^2$ or Tr $\boldsymbol{A}^2$ can be expressed in terms I_{A1} and I_{A2}, and Tr $\boldsymbol{A}^3$ can be expressed in terms of I_{A1}, I_{A2}, and I_{A3}, any polynomial in $\boldsymbol{A}$ that is invariant under arbitrary rotation has the functional form

$$\psi = \psi(\mathrm{Tr}\, \boldsymbol{A}, \mathrm{Tr}\, \boldsymbol{A}^2, \mathrm{Tr}\, \boldsymbol{A}^3). \tag{7.36}$$

These three traces form an irreducible basis for isotropic functions.

With isotropic functions depending on two tensors, $\boldsymbol{A}$ and $\boldsymbol{B}$, the irreducible basis consists of

$$\begin{gathered}\mathrm{Tr}\, \boldsymbol{A}, \mathrm{Tr}\, \boldsymbol{A}^2, \mathrm{Tr}\, \boldsymbol{A}^3,\\ \mathrm{Tr}\, \boldsymbol{B}, \mathrm{Tr}\, \boldsymbol{B}^2, \mathrm{Tr}\, \boldsymbol{B}^3,\\ \mathrm{Tr}\,(\boldsymbol{A} \cdot \boldsymbol{B}), \mathrm{Tr}\,(\boldsymbol{A} \cdot \boldsymbol{B}^2), \mathrm{Tr}(\boldsymbol{A}^2 \cdot \boldsymbol{B}), \mathrm{Tr}(\boldsymbol{A}^2 \cdot \boldsymbol{B}^2).\end{gathered} \tag{7.37}$$

Results for tensor-valued functions that exhibit various symmetries such as orthotropy, transverse isotropy, etc., have been worked out by Smith and Rivlin (1964), Adkins (1960), Spencer and Rivlin (1959), and Rivlin and Smith (1969).

SUGGESTED READING

Adkins, J. E. (1960). Symmetry relations for orthotropic and transversely isotropic materials, *Arch. Ration. Mech. Anal.*, **4**, 193–213.

Dahler, J. S. and Scriven, L. E. (1963). Theory of structured continua. I. General consideration of angular momentum and polarization, *Proc. R. Soc. London Ser. A*, **275**, 504–527.

Jaunzemis, W. (1967). *Continuum Mechanics*, Macmillan.

Rivlin, R. S. and Smith, G. F. (1969). Orthogonal integrity bases for *N* symmetric tensors, in *Contributions to Mechanics*, (S. Abir, ed.), Pergamon.

Smith, G. F. and Rivlin, R. S. (1964). Integrity bases for vectors. The crystal classes, *Arch. Ration. Mech. Anal.*, **15**, 169–221.

Spencer, A. J. M. and Rivlin, R. S. (1959). Finite integrity bases for five or fewer 3 × 3 matrices, *Arch. Ration. Mech. Anal.*, **2**, 435–446.

VanArsdale, W. E. (2003). Objective spin and the Rivlin–Ericksen model, *Acta Mech.*, **162**, 111–124.

Wedgewood, L. E. (1999). An objective rotation tensor applied to non-Newtonian fluid mechanics, *Rheol. Acta*, **38**, 91–99.

Wedgewood, L. E. and Geurts, K. R. (1995). A non-affine network model for polymer melts, *Rheol. Acta*, **34**, 196–208.

Zhou, X. and Tamma, K. K. (2003). On the applicability and stress update formulations for corotational stress rate hypoelasticity constitutive models, *Finite Elem. Anal. Design*, **39**, 783–816.

EXERCISES

7.1. For a fixed volume in space (control volume), we have material flowing in and out of its surface. Obtain the conservation of mass in this Eulerian description.

7.2. For the preceding problem, obtain the balance of linear and angular momentum.

7.3. Extend this approach to the balance of energy and the entropy production inequality.

7.4. Evaluate the suitability of the following rates for use as objective quantities:

(a)

$$\dot{A}_{ij} - A_{ik}\partial_k v_j - A_{jk}\partial_k v_i,$$

(b)

$$\dot{A}_{ij} - w_{ik}A_{kj} + A_{ik}w_{kj},$$

(c)

$$\dot{A}_{ij} - A_{kj}\partial_i v_k - A_{ik}\partial_k v_j,$$

(d)

$$\dot{A}_{ij} + A_{kj}\partial_i v_k + A_{ik}\partial_k v_j,$$

where A_{ij} is an objective tensor, $\partial_i v_j$ is the velocity gradient, and w_{ij} is the spin tensor.

7.5. Consider a system of particles $P_i (i = 1, 2, \ldots, N)$. The particle P_i with mass m_i is subjected to an external force $\boldsymbol{F}_i$. There are also interparticle forces $\boldsymbol{f}_{ij}$ on P_i that are due to P_j. Writing the second law of Newton for the motion of each particle, deduce the balance of linear and angular momentum for the system of particles.

8 Stress Tensor

We begin this chapter by recalling external forces and moments acting on a material continuum. Generally, these forces and moments may be grouped into body forces and body moments and contact forces and contact moments. The body forces and body moments are those exerted by external agencies such as gravity and magnetic fields on the body. They depend on the density of the material. The contact forces and contact moments are applied through the surface area of the body by other bodies. In elementary strength of materials, body moments are usually neglected.

8.1 External Forces and Moments

With the subscripts B for body and C for contact, we have

$$\boldsymbol{F} = \boldsymbol{F}_B + \boldsymbol{F}_C, \quad \boldsymbol{M} = \boldsymbol{M}_B + \boldsymbol{M}_C. \tag{8.1}$$

Introducing body force density $\boldsymbol{f}$ and moment density $\boldsymbol{\ell}$, we have

$$\boldsymbol{F}_B = \int_V \boldsymbol{f} \rho dv, \quad \boldsymbol{M}_B = \int_V \boldsymbol{\ell} \rho dv, \tag{8.2}$$

where ρ is the mass density and V is the current volume.

Similarly, the contact force and moment can be expressed as

$$\boldsymbol{F}_C = \int_S \boldsymbol{\sigma}^{(n)} da, \quad \boldsymbol{M}_C = \int_S \boldsymbol{m}^{(n)} da, \tag{8.3}$$

where $\boldsymbol{\sigma}^{(n)}$ is the surface traction, $\boldsymbol{m}^{(n)}$ is the surface moment on a unit area with normal $\boldsymbol{n}$. Again, in elementary courses, the surface moments are neglected. The total external moment has to be calculated including the moment that is due to the external forces.

8.2 Internal Forces and Moments

We introduce an imaginary surface cutting the body into two parts: B_I and B_{II} (see Fig. 8.1). At any point on this surface the outward normal of B_I is the negative of

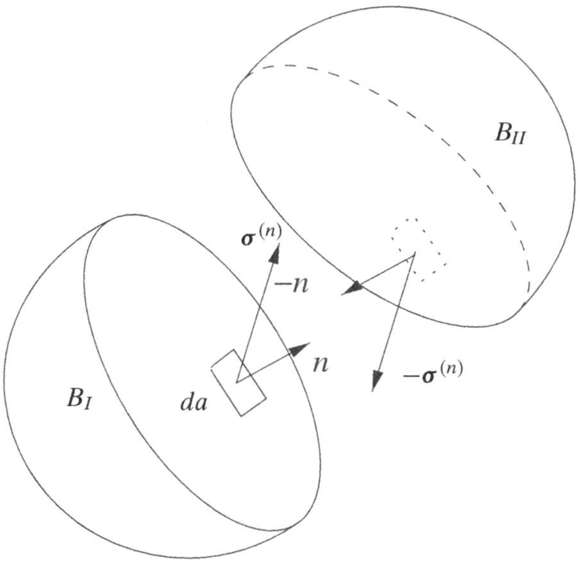

Figure 8.1. Traction vector on a plane.

the outward normal of B_{II}:

$$\boldsymbol{n}_I = -\boldsymbol{n}_{II}. \tag{8.4}$$

Concentrating on $B_{\rm I}$, the traction and surface moment acting on the newly created surface at a point $\boldsymbol{x}$ are $\boldsymbol{\sigma}^{(n)}$ and $\boldsymbol{m}^{(n)}$. These are called internal tractions and surface moments, respectively. These have dimensions of force per unit area and moment per unit area, respectively.

The traction acting on the same area of B_{II} with normal $-\boldsymbol{n}$ is

$$\boldsymbol{\sigma}^{(-n)} = -\boldsymbol{\sigma}^{(n)}. \tag{8.5}$$

Of all the possible surfaces passing through $\boldsymbol{x}$, we distinguish three coordinate planes with normals $\boldsymbol{e}_i$ and denote the tractions on these planes by $\boldsymbol{\sigma}^{(i)}$ and surface moments by $\boldsymbol{m}^{(i)}$. To relate the traction vectors on the coordinate planes passing through $\boldsymbol{x}$ and the traction vector $\boldsymbol{\sigma}^{(n)}$ acting on an arbitrary plane with normal $\boldsymbol{n}$, we construct an infinitesimal tetrahedron, as shown in Fig. 8.2, with the inclined plane with normal $\boldsymbol{n}$ having an area Δa. The areas Δa_i that are the projections of the area Δa along the axes x_i are related to Δa in the form

$$\boldsymbol{n}\Delta a = \boldsymbol{e}_i \Delta a_i, \quad \Delta a_i = n_i \Delta a. \tag{8.6}$$

Denoting the volume of the tetrahedron by Δv, we may balance the linear momentum of the tetrahedron as

$$\boldsymbol{\sigma}^{(n)}\Delta a - \boldsymbol{\sigma}^{(i)}\Delta a_i + \rho \Delta v \boldsymbol{f} = \frac{D}{Dt}(\boldsymbol{v}\rho\Delta v), \tag{8.7}$$

where the velocity $\boldsymbol{v}$, body force intensity, and all the tractions have to be interpreted as mean values within the volume and the respective areas. When we divide this equation by Δa and let Δa go to zero, the terms containing Δv, being of a higher

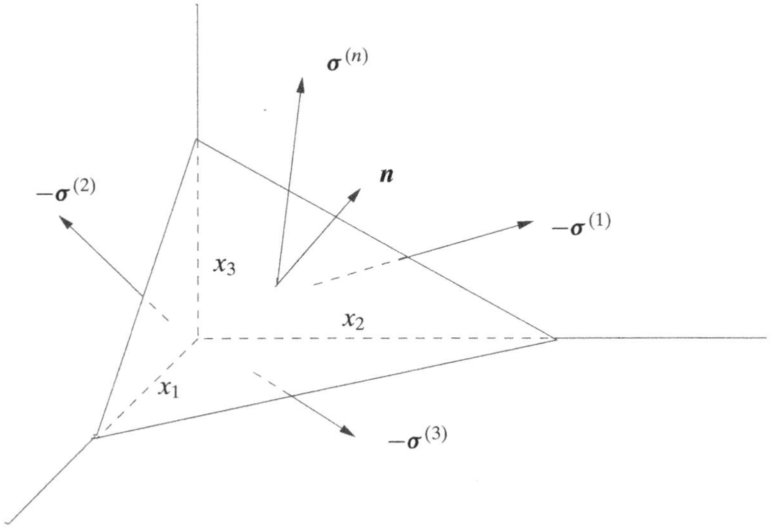

Figure 8.2. Traction vectors on the surfaces of a tetrahedron.

order in Δa, go to zero. The mean values become the corresponding values at the point $\boldsymbol{x}$, and we get

$$\boldsymbol{\sigma}^{(n)} = \boldsymbol{\sigma}^{(i)} n_i. \tag{8.8}$$

This result establishes the importance of the tractions on the three coordinate planes: Given the normal $\boldsymbol{n}$ of any plane, the traction $\boldsymbol{\sigma}^{(n)}$ can be found in terms of the tractions $\boldsymbol{\sigma}^{(i)}$.

Through the construction of a similar tetrahedron, using the balance of angular momentum, we can establish the surface moment relation:

$$\boldsymbol{m}^{(n)} = \boldsymbol{m}^{(i)} n_i. \tag{8.9}$$

In this process, we had to assume the angular momentum of the tetrahedron, net body moment, and the moments that are due to the body forces and surface tractions are of a higher order in Δa.

8.3 Cauchy Stress and Couple Stress Tensors

The traction vectors on the coordinate planes can be represented in terms of their components as

$$\boldsymbol{\sigma}^{(i)} = \sigma_{ij} \boldsymbol{e}_j, \tag{8.10}$$

where σ_{ij} are the components of a second-rank tensor, called the **Cauchy stress** tensor. The first index, i, represents the coordinate plane (in terms of normals), and the second index, j, represents the direction of the component. In matrix form we have

$$[\sigma_{ij}] = \begin{bmatrix} \sigma_{11} & \sigma_{12} & \sigma_{13} \\ \sigma_{21} & \sigma_{22} & \sigma_{23} \\ \sigma_{31} & \sigma_{32} & \sigma_{33} \end{bmatrix}. \tag{8.11}$$

The components σ_{11}, σ_{22}, and σ_{33} are called the normal stresses, and the remaining six off-diagonal components are called the shear stresses.

8.3.1 Transformation of the Stress Tensor

In Eq. (8.8), we may keep the plane fixed, maintaining the traction on it while a new coordinate system, x', is introduced. Denoting the stresses and the components of the normal $\boldsymbol{n}$ by σ'_{ij} and n'_i, respectively, in the new system, we get

$$\boldsymbol{\sigma}^{(n)} = \sigma'_{ij} n'_i \boldsymbol{e}'_j = \sigma_{lk} n_l \boldsymbol{e}_k. \tag{8.12}$$

Using $n_l = Q_{il} n'_i$ and $\boldsymbol{e}_k = Q_{jk} \boldsymbol{e}'_j$, we can establish the transformation law for the stress components as that for a second-rank tensor, namely,

$$\sigma'_{ij} = Q_{il} Q_{jk} \sigma_{lk}. \tag{8.13}$$

A similar representation holds for the surface moments:

$$\boldsymbol{m}^{(i)} = m_{ij} \boldsymbol{e}_j, \tag{8.14}$$

where m_{ij} are the components of the **couple stress** tensor. The couple stress tensor is also known as the Cosserat stress tensor in honor of the pioneering work on moment stresses carried out by the Cosserat brothers in the early 20th century. The transformation law for m_{ij} can be obtained to be that for a second-rank tensor, as we have done previously.

8.3.2 Principal Stresses

On an arbitrary plane with normal $\boldsymbol{n}$, the traction $\boldsymbol{\sigma}^{(n)}$ can be obtained as

$$\boldsymbol{\sigma}^{(n)} = \sigma_{ij} n_i \boldsymbol{e}_j. \tag{8.15}$$

We may resolve this with one component perpendicular to the plane and the remainder tangential to the plane. Denoting the perpendicular component by N, we have

$$N = \boldsymbol{n} \cdot \boldsymbol{\sigma}^{(n)} = \sigma_{ij} n_i n_j. \tag{8.16}$$

For a given stress tensor σ_{ij}, as the direction of the plane changes, the value of N changes according to the preceding relation. We call N the normal stress on the plane. It is important to know the maximum value of the normal stress and the corresponding plane when we design and analyze structures.

We should find the extremum value of N with respect to n_i with the constraint $n_i n_i = 1$. The procedure is identical to the one we used for extremum stretch, and it results in an eigenvalue problem. However, as we will see, the stress tensor is not symmetric when the body moments $\boldsymbol{\ell}$ are present. This fact appears to make this eigenvalue problem different from the case of the maximum stretch. But an inspection of the quadratic form in n_i shows that, because of the symmetry of $n_i n_j$, its coefficient is $\sigma_{ij} + \sigma_{ji}$. Thus we are free to use the symmetric form of the matrix σ_{ij} in computing the eigenvalues. Assuming the new matrix is the symmetric version $\sigma_{(ij)}$, the eigenvalue problem results in the system of equations

$$\sigma_{(ij)} n_j = \sigma n_i, \tag{8.17}$$

where σ is the eigenvalue and $\boldsymbol{n}$ is the eigenvector. For nontrivial solutions of this homogeneous system, we have to have

$$|\sigma_{(ij)} - \sigma \delta_{ij}| = 0. \tag{8.18}$$

Let $\sigma^{(k)}$ and $\boldsymbol{n}^{(k)}$ be the eigenvalues and eigenvectors of this problem. Multiplying Eq. (8.17) by n_i, we see that the eigenvalues are the stationary values of N, i.e.,

$$N = \sigma_{(ij)} n_j^{(k)} = \sigma^{(k)} n_i^{(k)} n_i^{(k)} = \sigma^{(k)}. \tag{8.19}$$

Again the three eigenvectors are mutually orthogonal, and these directions are called the principal directions and the corresponding normal tractions $\sigma^{(k)}$ are the principal stresses. The three invariants of the matrix $\sigma_{(ij)}$ are

$$\begin{aligned}
I_{\sigma 1} &= \sigma^{(1)} + \sigma^{(2)} + \sigma^{(3)} = \sigma_{(ii)}, \\
I_{\sigma 2} &= \sigma^{(1)}\sigma^{(2)} + \sigma^{(2)}\sigma^{(3)} + \sigma^{(3)}\sigma^{(1)} = \frac{1}{2}[\sigma_{(ii)}\sigma_{(jj)} - \sigma_{(ij)}\sigma_{(ij)}], \\
I_{\sigma 3} &= \sigma^{(1)}\sigma^{(2)}\sigma^{(3)} = |\sigma_{(ij)}|.
\end{aligned} \tag{8.20}$$

8.3.3 Shear Stress

The magnitude of the tangential component S of the traction vector on any plane may be computed as

$$S^2 = \boldsymbol{\sigma}^{(n)} \cdot \boldsymbol{\sigma}^{(n)} - (\boldsymbol{n} \cdot \boldsymbol{\sigma}^{(n)})^2. \tag{8.21}$$

On the principal planes $\boldsymbol{n}^{(k)}$, we have $\boldsymbol{\sigma}^{(n)} = \boldsymbol{n}^{(k)} \sigma^{(k)}$ and

$$S^2 = 0. \tag{8.22}$$

Thus there is no shear stress on the principal planes.

To obtain the planes on which the shear stress is stationary, it is convenient to use a new coordinate system with its axes oriented along the principal directions. In this system the stress matrix can be written as

$$[\sigma_{ij}] = \begin{bmatrix} \sigma_1 & 0 & 0 \\ 0 & \sigma_2 & 0 \\ 0 & 0 & \sigma_3 \end{bmatrix}, \tag{8.23}$$

where we have used the notation $\sigma_i = \sigma^{(i)}$. On any arbitrary plane with normal $\boldsymbol{n}$, the traction vector has the simple expression

$$\boldsymbol{\sigma}^{(n)} = n_1\sigma_1\boldsymbol{e}_1 + n_2\sigma_2\boldsymbol{e}_2 + n_3\sigma_3\boldsymbol{e}_3. \tag{8.24}$$

The shear stress on this plane becomes

$$S^2 = n_i^2\sigma_i^2 - (n_i\sigma_i)^2. \tag{8.25}$$

Adding the constraint $n_i n_i = 1$ and using a Lagrange multiplier Λ, we set up a function:

$$F(n_1, n_2, n_3) = n_i^2\sigma_i^2 - (n_i\sigma_i)^2 - \Lambda(n_i n_i - 1). \tag{8.26}$$

This function becomes stationary when $\partial F/\partial n_i = 0$. This way, we obtain the three equations

$$\begin{aligned}
\sigma_1^2 n_1 - 2H\sigma_1 n_1 - \Lambda n_1 &= 0,\\
\sigma_2^2 n_2 - 2H\sigma_2 n_2 - \Lambda n_2 &= 0,\\
\sigma_3^2 n_3 - 2H\sigma_3 n_3 - \Lambda n_3 &= 0,
\end{aligned} \tag{8.27}$$

where H is defined as

$$H = n_i\sigma_i.$$

We may verify that the three solutions of these equations are

$$\begin{aligned}
n_1 &= 0, \quad n_2^2 = n_3^2 = \frac{1}{2},\\
n_2 &= 0, \quad n_3^2 = n_1^2 = \frac{1}{2},\\
n_3 &= 0, \quad n_1^2 = n_2^2 = \frac{1}{2},
\end{aligned} \tag{8.28}$$

and the values of the stationary shear stresses for these three cases, respectively, are

$$S = \frac{|\sigma_2 - \sigma_3|}{2}, \frac{|\sigma_3 - \sigma_1|}{2}, \frac{|\sigma_1 - \sigma_2|}{2}. \tag{8.29}$$

8.3.4 Hydrostatic Pressure and Deviatoric Stresses

It is known that many materials respond differently to hydrostatic pressure than to shear stresses. When the normal stresses on the faces of an infinitesimal cube differ from each other, the hydrostatic pressure is defined as

$$-p = \frac{1}{3}\sigma_{ii} = \frac{1}{3}\left(\sigma^{(1)} + \sigma^{(2)} + \sigma^{(3)}\right). \tag{8.30}$$

If we remove the hydrostatic part from the stress tensor, what remains is called the deviatoric stress $\boldsymbol{s}$. Thus,

$$s_{ij} = \sigma_{ij} + p\delta_{ij} = \sigma_{ij} - \frac{1}{3}\sigma_{kk}\delta_{ij}. \tag{8.31}$$

8.3.5 Objective Stress Rates

There are materials that respond not only to the stress but also to the rate of change of stress in time. For a stress rate to be an admissible variable in a constitutive law, it has to be objective (frame indifferent). The obvious candidates such as $\partial\sigma_{ij}/\partial t$ and $D\sigma_{ij}/Dt$ can be ruled out if we use the appropriate rate of the relation

$$\sigma'_{ij} = Q_{im}Q_{jn}\sigma_{mn} \tag{8.32}$$

when the rotation matrix $\boldsymbol{Q}$ is time dependent.

The apparent rate of change in stress is best illustrated if we consider a bar under a constant stress σ_0. If we use a rotating frame with the x_1 axis along the axis of the bar at time $t = 0$, the frame shows $\sigma_{11} = \sigma_0$. As the frame rotates, we see σ_{11} diminishing, and, eventually, after a 90° rotation, σ_{22} has the value σ_0. During all this time, the bar itself did not experience any change in stress.

As it happens, an objective stress rate is not unique; a number of candidates have been proposed by various authors. A stress rate introduced by Jaumann in 1911, which is called the Jaumann stress rate or corotational stress rate, appears often in constitutive relations. The physical meaning of objective stress rates may be explored in the following way. At an arbitrary point P in a continuum, let us establish the origin of two coordinate systems: one, the Eulerian system with unit vectors $\boldsymbol{e}_i$ and coordinates x_i, and the second, a rotating system with unit vectors $\boldsymbol{e}'_i$ and coordinates x'_i. At time t, the two systems overlap. We assume the second system rotates with an angular velocity $\boldsymbol{\Omega}$. We will select $\boldsymbol{\Omega}$ later. Our goal is to compute the time rate of change of quantities from the point of view of an observer sitting on the rotating system. In an incremental time Δt, the base vectors of the rotating system are related to those of the fixed system as

$$\boldsymbol{e}'_i = P_{ij}(t')\boldsymbol{e}_i, \tag{8.33}$$

where $\boldsymbol{P}$ is an orthogonal matrix. When $t' = t$, the systems coincide and

$$P_{ij}(t) = \delta_{ij}. \tag{8.34}$$

The time rate of change of the base vectors $\boldsymbol{e}'_i$ can be expressed as

$$\dot{\boldsymbol{e}}'_i = -W_{ij}(t')\boldsymbol{e}'_j, \tag{8.35}$$

where $\boldsymbol{W}$ is a spin tensor related to $\boldsymbol{\Omega}$ through

$$W_{ij} = e_{ijk}\Omega_k. \tag{8.36}$$

Taking the material derivative of Eq. (8.33), we find

$$\dot{\boldsymbol{e}}'_i = \dot{P}_{ik}(t')\boldsymbol{e}_k = \dot{P}_{ik}P_{jk}\boldsymbol{e}'_j. \tag{8.37}$$

From this we observe that the spin tensor is related to $\dot{\boldsymbol{P}}$ in the form

$$\boldsymbol{W}(t') = -\dot{\boldsymbol{P}}\boldsymbol{P}^T, \quad \boldsymbol{W}(t) = -\dot{\boldsymbol{P}}. \tag{8.38}$$

When $t' = t + \Delta t$,

$$\boldsymbol{P}(t') = \boldsymbol{I} + \dot{\boldsymbol{P}}\Delta t = \boldsymbol{I} - \boldsymbol{W}\Delta t. \tag{8.39}$$

Any second-rank tensor field A_{ij} that is time dependent has components $A'_{ij}(t + \Delta t)$ in the rotating system. We define its rate of change as observed in the rotating system as

$$\hat{A}_{ij} \equiv \lim_{\Delta t \to 0} \frac{1}{\Delta t}[A'_{ij}(t + \Delta t) - A'_{ij}(t)]. \tag{8.40}$$

As the two coordinate systems coincide at time t,

$$A'_{ij}(t) = A_{ij}. \tag{8.41}$$

Now A_{ij} at $t + \Delta t$ is given by

$$A_{ij}(t + \Delta t) = A_{ij} + \dot{A}_{ij}\Delta t. \tag{8.42}$$

Using the transformation of coordinates and neglecting quadratic terms in Δt,

$$\begin{aligned} A'_{ij}(t + \Delta t) &= P_{im}P_{jn}A_{mn}(t + \Delta t) \\ &= (\delta_{im} - W_{im}\Delta t)(\delta_{jn} - W_{jn}\Delta t)(A_{mn} + \dot{A}_{mn}\Delta t) \\ &= A_{ij} + [\dot{A}_{ij} - W_{im}A_{mj} + A_{in}W_{nj}]\Delta t, \\ \hat{A}_{ij} &= \dot{A}_{ij} - W_{im}A_{mj} + A_{in}W_{nj}, \\ \hat{\boldsymbol{A}} &= \dot{\boldsymbol{A}} - \boldsymbol{W}\boldsymbol{A} + \boldsymbol{A}\boldsymbol{W}. \end{aligned} \tag{8.43}$$

Having the expression for the rate of change of a tensor in a rotating coordinate system, we would like to see how this quantity transforms under an arbitrary time-dependent rotation $\boldsymbol{Q}$ such that $x'_i = Q_{ij}(t)x_j$.

Starting with $\boldsymbol{A}' = \boldsymbol{Q}\boldsymbol{A}\boldsymbol{Q}^T$,

$$\begin{aligned} \hat{\boldsymbol{A}}' &= \dot{\boldsymbol{A}}' - \boldsymbol{W}'\boldsymbol{A}' + \boldsymbol{A}'\boldsymbol{W}' \\ &= \boldsymbol{Q}\dot{\boldsymbol{A}}\boldsymbol{Q}^T + \dot{\boldsymbol{Q}}\boldsymbol{A}\boldsymbol{Q}^T + \boldsymbol{Q}\boldsymbol{A}\dot{\boldsymbol{Q}}^T - \boldsymbol{W}'\boldsymbol{Q}\boldsymbol{A}\boldsymbol{Q}^T + \boldsymbol{Q}\boldsymbol{A}\boldsymbol{Q}^T\boldsymbol{W}' \\ &= \boldsymbol{Q}\,[\dot{\boldsymbol{A}} - \boldsymbol{Q}^T\boldsymbol{W}'\boldsymbol{Q}\boldsymbol{A} + \boldsymbol{A}\boldsymbol{Q}^T\boldsymbol{W}'\boldsymbol{Q} + \boldsymbol{Q}^T\dot{\boldsymbol{Q}}\boldsymbol{A} + \boldsymbol{A}\dot{\boldsymbol{Q}}^T\boldsymbol{Q}]\,\boldsymbol{Q}^T \\ &= \boldsymbol{Q}\,[\dot{\boldsymbol{A}} - \boldsymbol{W}\boldsymbol{A} + \boldsymbol{A}\boldsymbol{W}]\,\boldsymbol{Q}^T, \end{aligned} \tag{8.44}$$

where the last equation requires

$$\boldsymbol{W} = \boldsymbol{Q}^T\boldsymbol{W}'\boldsymbol{Q} + \boldsymbol{Q}^T\dot{\boldsymbol{Q}}. \tag{8.45}$$

For various choices of $\boldsymbol{W}$ satisfying the preceding condition, we obtain the corresponding objective stress rates $\hat{\boldsymbol{\sigma}}$. In the Jaumann stress rate, $\boldsymbol{W}$ is selected as the material spin tensor,

$$\boldsymbol{W} = \boldsymbol{w}, \quad \boldsymbol{\Omega} = \frac{1}{2}\nabla \times \boldsymbol{v}, \tag{8.46}$$

and the corresponding objective Cauchy stress rate is the Jaumann rate

$$\hat{\boldsymbol{\sigma}} = \dot{\boldsymbol{\sigma}} - \boldsymbol{w}\boldsymbol{\sigma} + \boldsymbol{\sigma}\boldsymbol{w}. \tag{8.47}$$

If we choose the rotation of the frame identical to the rotation $\boldsymbol{R}$ in the polar decomposition $\boldsymbol{F} = \boldsymbol{V}\boldsymbol{R}$, we obtain

$$\boldsymbol{W} = \dot{\boldsymbol{R}}\boldsymbol{R}^T, \tag{8.48}$$

and the corresponding stress rate is known as the Green–McInnis–Naghdi stress rate.

The Truesdell stress rate defined by

$$\hat{\boldsymbol{\sigma}} = \dot{\boldsymbol{\sigma}} - \nabla\boldsymbol{v}\cdot\boldsymbol{\sigma} - \boldsymbol{\sigma}\cdot(\nabla\boldsymbol{v})^T + \boldsymbol{\sigma}\nabla\cdot\boldsymbol{v}, \tag{8.49}$$

and the Oldroyd stress rate defined by

$$\hat{\boldsymbol{\sigma}} = \dot{\boldsymbol{\sigma}} - \nabla\boldsymbol{v}\cdot\boldsymbol{\sigma} - \boldsymbol{\sigma}\cdot(\nabla\boldsymbol{v})^T, \tag{8.50}$$

are also well known in the literature as objective stress rates.

Zhou and Tamma (2003) have shown that the logarithmic spin rate $\boldsymbol{W}$, defined by the equation

$$\frac{D}{Dt}\ln\boldsymbol{U} - \boldsymbol{W}\ln\boldsymbol{U} + (\ln\boldsymbol{U})\boldsymbol{W} = \boldsymbol{d}, \tag{8.51}$$

where $\boldsymbol{U}$ is the right stretch tensor and $\boldsymbol{d}$ is the deformation rate, leads to an objective stress rate that has certain numerical advantages over other objective stress rates.

8.4 Local Conservation and Balance Laws

The global conservation and balance laws presented in Chapter 7 can be transformed into their local (differential equation) forms by use of the stress tensors. We use the Gauss theorem to achieve this transformation. The balance of energy will have to wait until we discuss some concepts from thermodynamics (what Truesdell calls thermostatics).

8.4.1 Conservation of Mass

The global conservation equation

$$\frac{D}{Dt}\int_V \rho dv = 0 \tag{8.52}$$

can be written as

$$\int_V [\dot{\rho}dv + \rho(dv)^\bullet] = 0. \tag{8.53}$$

From this, using $(dv)^\bullet = \partial_i v_i dv$, we have the local form:

$$\dot{\rho} + \rho\partial_i v_i = 0. \tag{8.54}$$

8.4.2 Balance of Linear Momentum

From the global balance relation

$$\frac{D}{Dt}\int_V \boldsymbol{v}\rho dv = \boldsymbol{F}, \tag{8.55}$$

using the conservation of mass and the expression for the force $\boldsymbol{F}$, we get

$$\begin{aligned}\int_V \dot{\boldsymbol{v}}\rho dv &= \int_S \boldsymbol{\sigma}^{(n)} da + \int_V \boldsymbol{f}\rho dv\\ &= \int_S n_i \boldsymbol{\sigma}^{(i)} da + \int_V \boldsymbol{f}\rho dv\\ &= \int_V \partial_i \boldsymbol{\sigma}^{(i)} dv + \int_V \boldsymbol{f}\rho dv\\ &= \int_V \left[\partial_i \boldsymbol{\sigma}^{(i)} + \boldsymbol{f}\rho\right] dv.\end{aligned} \tag{8.56}$$

As this equation holds for arbitrary volumes, the integrands must be equal. Thus,

$$\partial_i \boldsymbol{\sigma}^{(i)} + \rho \boldsymbol{f} = \rho \dot{\boldsymbol{v}}. \tag{8.57}$$

Using the Cauchy stress tensor $\boldsymbol{\sigma}$, we can also write this as

$$\nabla \cdot \boldsymbol{\sigma} + \rho \boldsymbol{f} = \rho \boldsymbol{a}, \tag{8.58}$$

where $\boldsymbol{a}$ is the acceleration. The expanded form of this tensor equation is

$$\begin{aligned}\partial_1\sigma_{11} + \partial_2\sigma_{21} + \partial_3\sigma_{31} + \rho f_1 &= \rho a_1,\\ \partial_1\sigma_{12} + \partial_2\sigma_{22} + \partial_3\sigma_{32} + \rho f_2 &= \rho a_2,\\ \partial_1\sigma_{13} + \partial_2\sigma_{23} + \partial_3\sigma_{33} + \rho f_3 &= \rho a_3.\end{aligned} \tag{8.59}$$

8.4.3 Balance of Moment of Momentum (Angular Momentum)

The global balance of moment of momentum can be written as

$$\frac{D}{Dt}\int_V [\boldsymbol{J}\cdot\boldsymbol{\mu} + \boldsymbol{x}\times\boldsymbol{v}]\rho dv = \boldsymbol{M}, \tag{8.60}$$

where we have included the intrinsic angular velocity $\boldsymbol{\mu}$, which is distinct from the local angular velocity $\boldsymbol{\Omega}$, with a moment of inertia tensor $\rho\boldsymbol{J}$. In many cases of practical interest, we may need to include anisotropic characteristics in this inertia (consider a spinning ellipsoid, as an example). The integral on the left-hand side can be reduced to

$$\int_V [\boldsymbol{\alpha} + \boldsymbol{v}\times\boldsymbol{v} + \boldsymbol{x}\times\dot{\boldsymbol{v}}]\rho dv, \tag{8.61}$$

where

$$\boldsymbol{\alpha} = \frac{D}{Dt}(\boldsymbol{J}\cdot\boldsymbol{\mu}). \tag{8.62}$$

The moment on the right-hand side can be written as

$$M = \int_S [m^{(n)} + x \times \sigma^{(n)}] da + \int_V [\ell + x \times f] \rho dv. \tag{8.63}$$

Introducing the moment stress vectors and Cauchy stress vectors, we get

$$M = \int_S [n_i m^{(i)} + n_i x \times \sigma^{(i)}] da + \int_V [\ell + x \times f] \rho dv. \tag{8.64}$$

Using the Gauss theorem, we transform the surface integral to a volume integral:

$$M = \int_V [\partial_i m^{(i)} + x \times \partial_i \sigma^{(i)} + e_i \times \sigma^{(i)}] dv + \int_V [\ell + x \times f] \rho dv. \tag{8.65}$$

When we equate the terms on the left-hand side to those on the right-hand side of the equal sign, the terms involving x cancel because of the balance of linear momentum. What remains is the relation

$$\partial_i m^{(i)} + \rho \ell + e_i \times \sigma^{(i)} = \rho \alpha \Leftrightarrow \nabla \cdot m + \rho \ell + e_i \times \sigma^{(i)} = \rho \alpha. \tag{8.66}$$

In terms of the components, this becomes

$$\begin{aligned} \partial_1 m_{11} + \partial_2 m_{21} + \partial_3 m_{31} + \rho \ell_1 + \sigma_{23} - \sigma_{32} &= \rho \alpha_1, \\ \partial_1 m_{12} + \partial_2 m_{22} + \partial_3 m_{32} + \rho \ell_2 + \sigma_{31} - \sigma_{13} &= \rho \alpha_2, \\ \partial_1 m_{13} + \partial_2 m_{23} + \partial_3 m_{33} + \rho \ell_3 + \sigma_{12} - \sigma_{21} &= \rho \alpha_3. \end{aligned}$$

These equations show that the Cauchy stress σ may not be symmetric when the body moment ℓ, the couple stress m, or the intrinsic spin μ, is present. If all of these new quantities are absent, we obtain the classical result

$$\sigma_{ij} = \sigma_{ji} \tag{8.67}$$

as a consequence of the balance of angular momentum. The balance of linear and angular momentum equations are also called the equations of motion.

8.5 Material Description of the Equations of Motion

The equations of motion developed in the previous section are in terms of the spatial coordinates as independent variables. All the stress components in these equations have dimensions of force or surface moment per unit area of the current configuration. We may say that the Cauchy stress is the "true" stress experienced by the body at a given time. The body forces and the body moments have the dimensions of force or moment per unit mass and, as such, they are easy to describe in terms of the material coordinates. There are two ways of introducing the so-called pseudostresses in a material description: (a) keeping the elemental force vector constant, we define the traction vector by using the undeformed area, or (b) we treat the elemental force vector as a material element and scale the area as in (a). In the latter description the elemental force vector stretches and rotates as the material deforms. These two definitions result in stress tensors called the first and second Piola–Kirchhoff stress tensors.

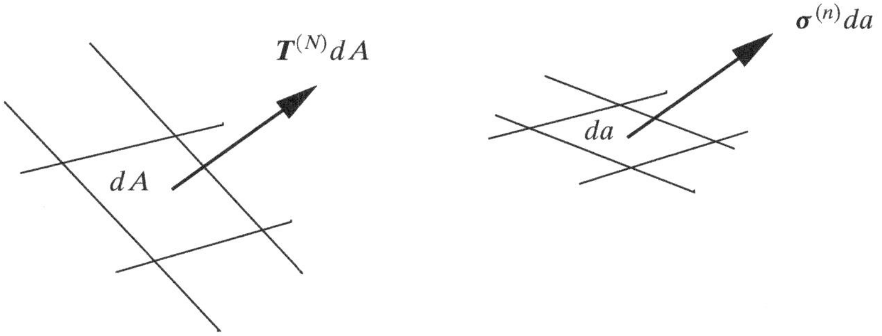

Figure 8.3. Traction vectors in the definition of the first Piola–Kirchhoff stress tensor.

8.5.1 First Piola–Kirchhoff Stress Tensor

We introduce a traction vector $\boldsymbol{T}^{(N)}$, which keeps the elemental force vector constant during deformation. As shown in Fig. 8.3,

$$\boldsymbol{T}^{(N)} dA = \boldsymbol{\sigma}^{(n)} da. \tag{8.68}$$

Here, $\boldsymbol{\sigma}^{(n)}$ is transported parallel to itself and scaled with the ratio of the areas to obtain $\boldsymbol{T}^{(N)}$. Using the tractions on the coordinate planes, we obtain

$$\begin{aligned} \boldsymbol{T}^{(\alpha)} N_\alpha dA &= \boldsymbol{\sigma}^{(i)} n_i da, \\ \boldsymbol{T}^{(\alpha)} dA_\alpha &= \boldsymbol{\sigma}^{(i)} da_i, \\ \boldsymbol{T}^{(\alpha)} dA_\alpha &= \boldsymbol{\sigma}^{(i)} J \partial_i X_\alpha dA_\alpha, \\ \boldsymbol{T}^{(\alpha)} &= \boldsymbol{\sigma}^{(i)} J \partial_i X_\alpha. \end{aligned} \tag{8.69}$$

The inverse form of this relation is

$$\boldsymbol{\sigma}^{(i)} = \boldsymbol{T}^{(\alpha)} j \partial_\alpha x_i. \tag{8.70}$$

Using the component form of these vectors,

$$\boldsymbol{T}^{(\alpha)} = T_{\alpha i} \boldsymbol{e}_i, \quad \boldsymbol{\sigma}^{(i)} = \sigma_{ij} \boldsymbol{e}_j, \tag{8.71}$$

we find

$$T_{\alpha i} = J \partial_k X_\alpha \sigma_{ki}, \quad \sigma_{ij} = j \partial_\alpha x_i T_{\alpha j}. \tag{8.72}$$

The tensor $T_{\alpha i}$ is called the first Piola–Kirchhoff stress or the Lagrange stress tensor.

To rewrite the equations of motion, let us consider the three terms in the Cauchy equations:

$$\rho a_i = j \rho_0 \frac{\partial^2 x_i}{\partial t^2}, \quad \rho f_i = j \rho_0 F_i, \tag{8.73}$$

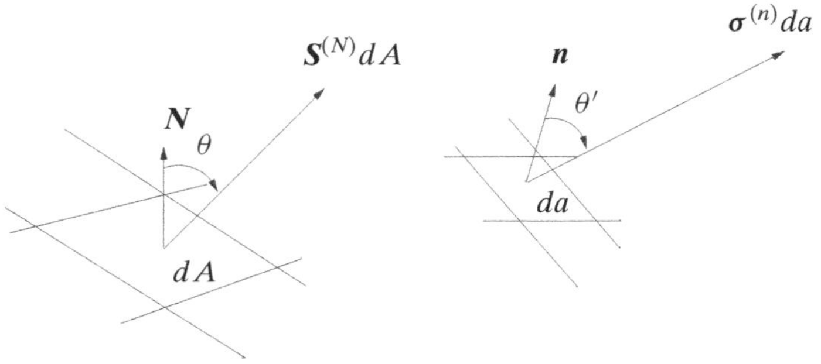

Figure 8.4. Traction vectors in the definition of the second Piola–Kirchhoff stress tensor.

where $F_i = f_i$;

$$\begin{aligned}\partial_k \sigma_{ki} &= \partial_k(j\partial_\alpha x_k T_{\alpha i}) \\ &= (\partial_k j)\partial_\alpha x_k T_{\alpha i} + j(\partial_k \partial_\alpha x_k)T_{\alpha i} + j\partial_\alpha x_k(\partial_k T_{\alpha i}) \\ &= j\partial_\alpha T_{\alpha i},\end{aligned} \tag{8.74}$$

where we have used

$$j = j(\partial_m X_\beta), \quad \partial_\alpha \partial_m X_\beta = \partial_m \partial_\alpha X_\beta = 0. \tag{8.75}$$

Combining the preceding three results and canceling the factor j, we find that the equations of motion in material coordinates come out as

$$\partial_\alpha T_{\alpha i} + \rho_0 F_i = \rho_0 \frac{\partial^2 x_i}{\partial t^2}. \tag{8.76}$$

8.5.2 Second Piola–Kirchhoff Stress Tensor

The first Piola–Kirchhoff stress tensor. We just defined may not be symmetric when the Cauchy stress is symmetric. The second Piola–Kirchhoff stress, also known as the Kirchhoff stress, which is symmetric when the Cauchy stress is symmetric, is defined as

$$S_{\alpha\beta} = J\partial_i X_\alpha \partial_j X_\beta \sigma_{ij} = T_{\alpha i}\partial_i X_\beta. \tag{8.77}$$

We may associate traction vectors on the coordinate planes

$$\boldsymbol{S}^{(\alpha)} = S_{\alpha\beta}\boldsymbol{e}_\beta = J\partial_i X_\alpha \partial_j X_\beta \sigma_{ij}\boldsymbol{e}_\beta = T_{\alpha i}\partial_i X_\beta \boldsymbol{e}_\beta, \tag{8.78}$$

and those on any plane

$$\boldsymbol{S}^{(N)} = \boldsymbol{S}^{(\alpha)} N_\alpha = J\partial_i X_\alpha \partial_j X_\beta \sigma_{ij} N_\alpha \boldsymbol{e}_\beta = \partial_i X_\beta T_{\alpha i} N_\alpha \boldsymbol{e}_\alpha = \boldsymbol{T}^{(N)} \cdot \boldsymbol{f}, \tag{8.79}$$

where $\boldsymbol{f}$ is the deformation gradient (not to be confused with the body force). In matrix notation we have

$$\boldsymbol{S} = \boldsymbol{T}\boldsymbol{f} = J\boldsymbol{f}^T \boldsymbol{\sigma}\boldsymbol{f}. \tag{8.80}$$

The last expression clearly shows the symmetry of $\boldsymbol{S}$ when $\boldsymbol{\sigma}$ is symmetric.

A physical interpretation of the traction $\boldsymbol{S}^{(N)}$ may be obtained as follows. Consider a material element $\boldsymbol{D}$ attached to the area dA in the undeformed configuration. In the current configuration,

$$\boldsymbol{D} = D_\alpha \boldsymbol{e}_\alpha = dX_\alpha \boldsymbol{e}_\alpha = dx_i \partial_i X_\alpha \boldsymbol{e}_\alpha = \boldsymbol{d} \cdot \boldsymbol{f}. \tag{8.81}$$

Comparing this with

$$\boldsymbol{S}^{(N)} = \boldsymbol{T}^{(N)} \cdot \boldsymbol{f} \tag{8.82}$$

shows that the Kirchhoff traction vector deforms as a material element to end up in the current configuration as the Lagrange traction vector. This is shown in Fig. 8.4. Because the force vector in the undeformed configuration is treated as a material element, the angles θ and θ' are not the same. Angles between material elements may get distorted due to shearing.

The equation of motion can be written in terms of $\boldsymbol{S}$ as

$$\partial_\alpha (S_{\alpha\beta} \partial_\beta x_i) + \rho_0 F_i = \rho_0 \frac{\partial^2 x_i}{\partial t^2}. \tag{8.83}$$

Because of the deformation gradient multiplying the stress tensor, these equations are nonlinear for finite deformations. Using $x_1 = X_1 + U_1$, etc., we can rewrite these as

$$\partial_\alpha [S_{\alpha\beta}(\delta_{\beta\gamma} + \partial_\beta U_\gamma)] + \rho_0 F_\gamma = \rho_0 \ddot{U}_\gamma. \tag{8.84}$$

The material description of the angular momentum balance in terms of couple stresses can be carried out in a similar fashion.

SUGGESTED READING

Zhou, X. and Tamma, K. K. (2003). On the applicability and stress update formulations for corotational stress rate hypoelasticity constitutive models, *Finite Elem. Anal. Design*, **39**, 783–816.

EXERCISES

8.1. At a point in a continuum, the stress matrix is given by

$$\boldsymbol{\sigma} = \begin{bmatrix} 4 & 3 & 0 \\ 0 & -4 & 0 \\ 0 & 0 & 2 \end{bmatrix}.$$

(a) Obtain the traction vector on a plane with normal vector $(2/3, -2/3, 1/3)$.
(b) Obtain the normal and tangential components of the traction on this plane.

8.2. The stress matrix at a point is given as

$$\boldsymbol{\sigma} = \begin{bmatrix} 0 & 2 & 1 \\ 2 & \sigma_{22} & 2 \\ 1 & 2 & 0 \end{bmatrix}.$$

(a) Obtain the value of σ_{22} if there is a traction-free plane passing through this point.
(b) Determine the unit normal to this plane.

8.3. In a medium supporting couple stresses, the Cauchy stress tensor is given as

$$\boldsymbol{\sigma} = \begin{bmatrix} 2 & 1 & 0 \\ -3 & 2 & 0 \\ 0 & 0 & 3 \end{bmatrix}.$$

Obtain the principal stresses and their directions.

8.4. A body is in **plane stress** if $\sigma_{3i} = 0$. In addition, if the body force $\boldsymbol{f} = \boldsymbol{0}$ and the acceleration $\boldsymbol{a} = \boldsymbol{0}$, show that

$$\sigma_{11} = \phi_{,22}, \quad \sigma_{22} = \phi_{,11}, \quad \sigma_{12} = \sigma_{21} = -\phi_{,12}$$

satisfy the equations of motion. The function ϕ is called the Airy stress function. Obtain the traction vector on a boundary $s = s(x_1, x_2)$ in terms of ϕ.

8.5. Obtain the principal stresses and their planes for

$$\boldsymbol{\sigma} = \begin{bmatrix} 1 & 0 & 0 \\ 0 & 4 & -3 \\ 0 & -3 & 4 \end{bmatrix}.$$

What are the three invariants of this matrix?

8.6. Show that the second invariant of the deviatoric stress, s_{ij}, can be written as

(a)

$$I_{s2} = s_{11}s_{22} + s_{22}s_{33} + s_{33}s_{11} - (s_{12}^2 + s_{23}^2 + s_{31}^2),$$

(b)

$$I_{s2} = \frac{1}{2}(s_{ii}s_{jj} - s_{ij}s_{ij}),$$

(c)

$$I_{s2} = -\frac{1}{6}\left[(\sigma_{11} - \sigma_{22})^2 + (\sigma_{22} - \sigma_{33})^2 + (\sigma_{33} - \sigma_{11})^2\right] - (\sigma_{12}^2 + \sigma_{23}^2 + \sigma_{31}^2).$$

8.7. For the deformation

$$x_1 = X_1 + kX_2, \quad x_2 = X_2, \quad x_3 = X_3,$$

the Cauchy stress at a point is given as

$$\boldsymbol{\sigma} = \begin{bmatrix} 0 & 1 & 2 \\ 1 & 0 & 2 \\ 2 & 2 & 0 \end{bmatrix}.$$

(a) Obtain the components of the first and second Piola–Kirchhoff stresses.
(b) Obtain the traction vector on a plane in the current configuration with normal $\boldsymbol{n} = (1, 1, 0)/\sqrt{2}$.
(c) Show corresponding traction vectors of the two Piola–Kirchhoff stresses.

8.8. Show that the Oldroyd stress rate and the Truesdell stress rate are objective.

8.9. Consider a simple shear flow, in which

$$v_1 = \gamma x_2, \quad v_2 = 0, \quad v_3 = 0,$$

and γ is a constant. Obtain the material spin tensor $\boldsymbol{w}$. Also obtain the components of the objective Jaumann stress rate.

8.10. For a hyperelastic material there is a strain energy density function defined by

$$\rho u = \frac{1}{2}\sigma_{ij}e_{ij}$$

per unit volume, and the total strain energy is given by

$$U = \int_V \rho u dv.$$

In the material description the strain energy density is defined as

$$\rho_0 u_0 = \frac{1}{2}\Pi_{\alpha\beta}E_{\alpha\beta}$$

per unit volume, where $\Pi_{\alpha\beta}$ is a symmetric stress measure. Comparing the total strain energy, obtain an expression for the stress measure $\Pi_{\alpha\beta}$ and relate it to the Kirchhoff stress.

9 Energy and Entropy Constraints

In this chapter we discuss the constraints imposed on the mathematical formulations of material behavior by the balance of energy and the irreversible part of the entropy production. In other words, we discuss the implications of the first and second laws of thermodynamics on the constitutive relations of continuum mechanics.

9.1 Classical Thermodynamics

In classical thermodynamics, we call a collection of particles under consideration a system. Let us denote this system by $\mathcal{B}$, and let $\mathcal{S}$ be the surface enclosing it. We assume $\mathcal{B}$ is isolated, meaning there is no mass moving across the surface $\mathcal{S}$. By moving this surface, we can do work on the system. By bringing this surface into contact with another surface belonging to a different system, heat energy can be made to flow into the system or out of the system. For basic properties of the system, we have the extensive variables: V, the volume; E, the internal energy; and S, the entropy. We also have intensive variables: p, the pressure; T, the temperature; and the chemical potential, etc. In continuum mechanics, we need to extend the idea of a pressure p to the stress tensor $\boldsymbol{\sigma}$ and the idea of a specific volume to an appropriate strain tensor. If the surface $\mathcal{S}$ is assumed to prevent heat flow across it, the system is called adiabatic.

When all the information needed to characterize a system (to the extent we require) is known, we say we know the state of the system. The properties of the system in a particular state are called state variables. For example, in the case of an ideal gas, we would like to know its volume, pressure, temperature, energy content, specific volume, and entropy content. These form the state variables. We do not care to know the velocities and locations of the zillions of individual molecules, and we do not include in the energy content the rest mass in the relativistic sense, as this does not change in the processes we have in mind. The processes we have in mind involve taking the system from one state to another, in a slow fashion. The system is said to be in equilibrium if the state variables have time-independent values. In our example of an ideal gas, if the volume is suddenly changed by ΔV by pushing the surface, we can expect certain wave motions inside that would eventually decay and the system will reach a new state of equilibrium.

If certain state variables can be expressed as functions of certain other state variables, such functions are called equations of states. If the state variables are independent of the space coordinates, the system is called homogeneous. If two systems are brought into contact and if the surfaces are allowed to conduct heat, one of the state variables, namely the temperature, assumes a common value after sufficient time.

9.2 Balance of Energy

The material continuum is endowed with two forms of energy–the kinetic energy, K, and the internal energy, E. We allow the flow of energy **into the body** in the forms of mechanical power P and heat power H. We briefly discuss other forms of energy and power input at the end of this section.

The kinetic energy K and the internal energy E are defined as

$$K = \frac{1}{2}\int_V [\boldsymbol{v} \cdot \boldsymbol{v} + \boldsymbol{\mu} \cdot \boldsymbol{J} \cdot \boldsymbol{\mu}]\rho dv, \qquad E = \int_V \varepsilon \rho dv. \tag{9.1}$$

In the first of Eqs. (9.1), the kinetic energy density consists of the classical $\rho v^2/2$ and a "microstructural" contribution $\rho J \mu^2/2$, where ρJ represents the microstructural moment of inertia. Further, we have introduced a certain anisotropy for the microstructure and written the rotational energy in the tensor form by using the moment of inertia tensor $\rho \boldsymbol{J}$. The vector $\boldsymbol{\mu}$ is an independent kinematic variable representing the microstructural angular velocity, which is not related to the local spin of classical continua. It is useful to think of steam, which is composed of H_2O molecules, as an example. These molecules have a common structure with an associated inertia tensor, but they may have independent angular velocities (spins). Of course, we are not going to count individual molecules and their properties. Other examples are liquid crystals and granular materials (see Cowin, 1978; Christoffersen, Mehrabadi, and Nemat-Nasser, 1981). The global conservation of energy reads

$$\frac{D}{Dt}(K + E) = P + H. \tag{9.2}$$

The mechanical power input into the body consists of the rates of work done on the body by surface tractions, surface moments, body forces, and body moments:

$$P = \int_S [\boldsymbol{\sigma}^{(n)} \cdot \boldsymbol{v} + \boldsymbol{m}^{(n)} \cdot \boldsymbol{\mu}]da + \int_V [\boldsymbol{f} \cdot \boldsymbol{v} + \boldsymbol{\ell} \cdot \boldsymbol{\mu}]\rho dv, \tag{9.3}$$

where $\boldsymbol{\mu}$ is the angular velocity vector conjugate to the surface moment vector.

The heat power is due to the rate of heat flow into the body, $\boldsymbol{q}$, and the rate of heat generation inside the body, h:

$$H = \int_S \boldsymbol{q} \cdot \boldsymbol{n} da + \int_V h\rho dv. \tag{9.4}$$

The next step in our calculation is to obtain the local form of the conservation of energy. We begin with

$$\frac{DK}{Dt} = \frac{1}{2}\int_V \left[\frac{D}{Dt}(v_i v_i + J_{ij}\mu_i\mu_j)\right]\rho dv = \int_V [a_i v_i + \alpha_i \mu_i]\,\rho dv, \tag{9.5}$$

where we have used the definition

$$\alpha_i = \frac{D}{Dt}J_{ij}\mu_j$$

and the conservation of mass for simplification.

The rate of change of internal energy is obtained as

$$\frac{DE}{Dt} = \int_V \dot{\varepsilon}\rho dv. \tag{9.6}$$

Of the two forms of power input, first we consider the simpler expression,

$$H = \int_S q_i n_i da + \int_V h\rho dv = \int_V [\partial_i q_i + \rho h] dv. \tag{9.7}$$

In the expression for P,

$$\begin{aligned} \boldsymbol{\sigma}^{(n)} \cdot \boldsymbol{v} &= n_i \sigma_{ij} v_j, \\ \boldsymbol{m}^{(n)} \cdot \boldsymbol{\mu} &= n_i m_{ij}\mu_j. \end{aligned} \tag{9.8}$$

Using the Gauss theorem, we get

$$\begin{aligned} P &= \int_V [\partial_i(\sigma_{ij}v_j + m_{ij}\mu_j) + \rho(f_j v_j + \ell_j\mu_j)] dv \\ &= \int_V [\sigma_{ij}\partial_i v_j + m_{ij}\partial_i\mu_j + v_j(\partial_i\sigma_{ij} + \rho f_j) + \mu_j(\partial_i m_{ij} + \rho\ell_j)] dv. \end{aligned} \tag{9.9}$$

We introduce the relations

$$\begin{aligned} \partial_i\sigma_{ij} + \rho f_j &= \rho a_j, \\ \partial_i m_{ij} + \rho\ell_j &= \rho\alpha_j - e_{jkl}\sigma_{kl} \end{aligned}$$

in Eq. (9.9) to obtain the local form of the conservation of energy:

$$\rho\dot{\varepsilon} = \rho h + \partial_i q_i + \sigma_{ij}\partial_i v_j + m_{ij}\partial_i\mu_j - e_{jkl}\sigma_{kl}\mu_j. \tag{9.10}$$

In this energy conservation relation, the third and the fourth terms are called the stress power and the couple stress power, respectively. Further simplification of this relation is possible if (a) the couple stresses are absent and (b) part of the stress tensor can be derived from a potential.

9.3 Clausius–Duhem Inequality

In classical thermostatics, the incremental form of the internal energy is written as

$$\Delta E = \delta W + \delta Q, \tag{9.11}$$

where δW is the work done on the system and δQ is the heat added. We have reserved the symbol Δ to denote change in a state variable. The quantities W and Q are not state variables, and the symbol δ in front of them just denotes "small" increments. For our system, the increment of the state function, entropy, which is due to heat flow across the surface, is given by

$$\Delta S = \frac{\delta Q}{T}, \tag{9.12}$$

where T is the absolute temperature.

Substituting this in the energy equation, we find

$$\Delta E = \delta W + T \Delta S. \tag{9.13}$$

In irreversible thermodynamics, the entropy production ΔS consists of two parts. One is due to the entropy flow into the body, ΔS_E, which is associated with the heat flow, and the other, ΔS_I, is due to irreversible processes taking place inside the body. The total entropy S is considered to be a state variable, and Eq. (9.13) is taken as a fundamental relation, although its original derivation was based on heat flow only.

The work term in Eq. (9.13) is replaced with stresses τ_{ij} and thermodynamic moment stresses λ_{ij}, producing incremental work per unit volume because of the increments in a corresponding array of state variables. Without being specific, let ϵ_{ij} and ν_{ij} be these state variables. The work done by these stresses and moment stresses have to be reversible.

Introducing internal energy density ε and specific entropy s, we have the local form of the equation

$$\rho \Delta \varepsilon = \rho T \Delta s + \tau_{ij} \Delta \epsilon_{ij} + \lambda_{ij} \Delta \nu_{ij}. \tag{9.14}$$

This is a version of the Gibbs equation in thermodynamics. In the limit, the Gibbs equation gives the defining relations

$$T = \frac{\partial \varepsilon}{\partial s}, \quad \tau_{ij} = \rho \frac{\partial \varepsilon}{\partial \epsilon_{ij}}, \quad \lambda_{ij} = \rho \frac{\partial \varepsilon}{\partial \nu_{ij}}. \tag{9.15}$$

Further, dividing the Gibbs eqution by δt and taking the limit, we find

$$\rho \dot{s} = \frac{1}{T}(\rho \dot{\varepsilon} - \tau_{ij} \dot{\epsilon}_{ij} - \lambda_{ij} \dot{\nu}_{ij}). \tag{9.16}$$

Substituting energy equation (9.10) into Eq. (9.16), we find that Eq. (9.16) becomes

$$\rho \dot{s} = \frac{1}{T}(\rho h + \partial_i q_i + \sigma_{ij} \partial_i v_j + m_{ij} \partial_i \mu_j - e_{jkl} \sigma_{kl} \mu_j - \tau_{ij} \dot{\epsilon}_{ij} - \lambda_{ij} \dot{\nu}_{ij}). \tag{9.17}$$

At first glance, this equation appears to have too many variables! To reduce it to a simpler form, first we have to choose ϵ_{ij} and ν_{ij}, with their rates relating to $\partial_i v_j$ and $\partial_i \mu_j$. Second, parts of the stresses σ_{ij} and m_{ij} may contribute to reversible work, and we can relate them to τ_{ij} and λ_{ij}.

The rate of global entropy gain that is due to heat input is seen to be

$$\dot{S}_E = \int_S \frac{n_i q_i}{T} da + \int_V \frac{\rho h}{T} dv. \tag{9.18}$$

From this the local form is obtained as

$$\begin{aligned} \rho \dot{s}_E &= \partial_i \left(\frac{q_i}{T}\right) + \frac{\rho h}{T} \\ &= \frac{1}{T}(\partial_i q_i + \rho h) - \frac{q_i \partial_i T}{T^2}. \end{aligned} \tag{9.19}$$

Subtracting this from entropy equation (9.17), we get

$$\rho \dot{s}_I = \frac{1}{T}\left(\sigma_{ij}\partial_i v_j + m_{ij}\partial_i \mu_j - e_{jkl}\sigma_{kl}\mu_j - \tau_{ij}\dot{\epsilon}_{ij} - \lambda_{ij}\dot{\nu}_{ij} + \frac{q_i \partial_i T}{T}\right). \tag{9.20}$$

As already mentioned, further simplification is achieved if we assume that the Cauchy stress $\boldsymbol{\sigma}$ and the couple stress $\boldsymbol{m}$ can be resolved into a hyperelastic part that can be derived from a potential and an irreversible part that contributes to the internal entropy generation. The potential associated with the hyperelastic part is known as the strain energy density function, u. This potential is a function of the state variables ϵ_{ij} and ν_{ij}. This resolution is expressed as

$$\boldsymbol{\sigma} = \boldsymbol{\sigma}^0 + \boldsymbol{\sigma}^I, \quad \boldsymbol{m} = \boldsymbol{m}^0 + \boldsymbol{m}^I. \tag{9.21}$$

Then we choose τ_{ij} and λ_{ij} to satisfy

$$\sigma_{ij}^0(\partial_i v_j - e_{kij}\mu_k) + m_{ij}^0 \partial_i \mu_j = \tau_{ij}\dot{\epsilon}_{ij} + \lambda_{ij}\dot{\nu}_{ij} = \rho \dot{u}. \tag{9.22}$$

This leaves the internal entropy production rate:

$$\rho \dot{s}_I = \frac{1}{T}\left[\sigma_{ij}^I(\partial_i v_j - e_{kij}\mu_k) + m_{ij}^I \partial_i \mu_j + \frac{q_i \partial_i T}{T}\right]. \tag{9.23}$$

The Clausius–Duhem inequality, in its local form, states that

$$\dot{s}_I \geq 0, \tag{9.24}$$

with the equality satisfied if all processes are reversible. It has been shown by Prigogine, using statistical mechanics, that the entropy production rate we have derived is valid only for what are called "first-order" processes, in which the irreversible stresses and heat flux may be expressed as linear functions of the variables $\partial_i v_j$, $\partial_i \mu_j$, and $\partial_i T$. The coefficient matrix in such a system of relations must be positive semidefinite to satisfy the Clausius–Duhem inequality.

Let us illustrate this with two simple cases.

9.3.1 Fourier's Law of Heat Conduction

In the generalized Fourier's law for anisotropic materials, the heat flux can be expressed as

$$q_i = k_{ij}\partial_j T, \tag{9.25}$$

where k_{ij} is the conductivity matrix. If we assume a process in which there are no stresses, using this in the entropy production, we find

$$k_{ij}\partial_i T \partial_j T \geq 0 \tag{9.26}$$

for an arbitrary temperature gradient. For this to be true, the matrix $\boldsymbol{k}$ has to be positive semidefinite.

9.3.2 Newton's Law of Viscosity

We consider a fluid flowing over a plate with velocity parallel to the plate. In its simplest form, with the x coordinate parallel to the plate and the y coordinate perpendicular, the only velocity gradient is $\partial_y v_x$. Newton's law of viscosity states

$$\sigma_{xy} = \mu \partial_y v_x, \tag{9.27}$$

where μ is the viscosity. The Clausius–Duhem inequality requires that $\mu \geq 0$ if all other terms are absent.

9.3.3 Onsager's Principle

The rate of specific entropy production can be written in terms of what are called generalized forces X_k and fluxes J_k in the form

$$\rho \dot{s}_I = X_k J_k. \tag{9.28}$$

In Eq. (9.23), we can identify

$$X_k \Rightarrow (\sigma_{ij}^I, m_{ij}^I, \partial_i T/T),$$

$$J_k \Rightarrow (\partial_i v_j - e_{kij}\mu_k, \partial_i \mu_j, q_i)\frac{1}{T}. \tag{9.29}$$

At thermodynamic equilibrium, $\dot{s}_I = 0$, and all the fluxes and forces are also zero. Onsager has shown that, close to the equilibrium state, a system of linear relations exists that relates the fluxes and forces:

$$J_k = L_{km} X_m. \tag{9.30}$$

Fourier's law and Newton's law, previously discussed, are examples of such linear relations. Onsager's principle states that the matrix L_{km} is symmetric. Further, the Clausius–Duhem inequality requires L_{km} to be positive semidefinite.

9.3.4 Strain Energy Density

To simplify the form of the strain energy density function u, let us assume that, in Eqs. (9.21),

$$\boldsymbol{m} = \boldsymbol{0}, \quad \boldsymbol{\ell} = \boldsymbol{0}, \quad \boldsymbol{\mu} = \boldsymbol{0}, \tag{9.31}$$

which implies that the only stress remaining is the Cauchy stress $\boldsymbol{\sigma}$, which is now symmetric, satisfying the balance of angular momentum. Let us also assume that $\sigma^I_{ij} = 0$.

Let

$$u = u(E_{\alpha\beta}), \quad \epsilon_{ij} \Rightarrow E_{\alpha\beta}, \quad \nu_{ij} = 0. \tag{9.32}$$

Then

$$\begin{aligned} \rho \dot{u} &= \rho \frac{\partial u}{\partial E_{\alpha\beta}} \dot{E}_{\alpha\beta} \\ &= \rho \frac{\partial u}{\partial E_{\alpha\beta}} d_{ij} \partial_\alpha x_i \partial_\beta x_j. \end{aligned} \tag{9.33}$$

In Eq. (9.22), using $\partial_i v_j = d_{ij} + w_{ij}$ and $\sigma^0_{ij} w_{ij} = 0$, we find

$$\begin{aligned} \sigma^0_{ij} \partial_i X_\alpha \partial_j X_\beta &= \rho \frac{\partial u}{\partial E_{\alpha\beta}}, \\ S^0_{\alpha\beta} &= \rho_0 \frac{\partial u}{\partial E_{\alpha\beta}}. \end{aligned} \tag{9.34}$$

We conclude by noting that the second Piola–Kirchhoff stress $S_{\alpha\beta}$ is the stress measure τ_{ij} if $E_{\alpha\beta}$ is the state variable ϵ_{ij}. We say that the stress $\boldsymbol{S}^0$ is conjugate to the strain $\boldsymbol{E}$.

9.3.5 Ideal Gas

A measure of volume change is the volume ratio J (this is the Jacobian determinant, not to be confused with the Onsager fluxes). If we assume the strain energy has the form

$$u = u(J), \tag{9.35}$$

we find

$$\rho \frac{du}{dJ} \dot{J} = \sigma_{ij}(d_{ij} + w_{ij}) = \sigma_{ij} d_{ij}. \tag{9.36}$$

Using $\dot{J} = J d_{ii}$, we have

$$\rho_0 \frac{du}{dJ} d_{ij} \delta_{ij} = t_{ij} d_{ij}. \tag{9.37}$$

Thus the only stress sustainable in this medium is the hydrostatic pressure,

$$\rho_0 \frac{du}{dJ} \delta_{ij} = t_{ij}. \tag{9.38}$$

We see the well-known state variable from classical thermodynamics, namely, the pressure p, given by

$$p = -\rho_0 \frac{du}{dJ}, \tag{9.39}$$

emerging as the only stress in this system.

9.4 Internal Energy

For the sake of simplicity, let us consider a homogeneous system with the internal energy

$$E = E(S, \boldsymbol{\epsilon}), \tag{9.40}$$

where the group of variables $\boldsymbol{\epsilon}$ represents the specific volume in classical thermodynamics and the six strain components for the hyperelastic case previously considered. This group will contain more strain measures if microstructural spins are allowed. In applications involving chemical mixtures, various chemical concentrations have to be added to this group. For granular materials, the void fraction is an important state variable, which also has to be included. We may assume this group can be written as a linear array ϵ_i (we will not use the term vector, as our array does not transform according to the rules). We denote their conjugate variables, called thermodynamic tensions, by the array τ_i, and the inner product of these two arrays, $\tau_i \epsilon_i$, has the dimension of energy and is invariant under translations and rotations. Equation (9.40) is called a **caloric equation of state**. We invert this relation to have

$$S = S(E, \boldsymbol{\epsilon}). \tag{9.41}$$

At thermodynamic equilibrium, $\dot{S} = 0$, and for an adiabatic process there is no entropy flow into the body. Internal processes leading to equilibrium may generate entropy, and at equilibrium the entropy reaches a maximum. This is called the maximum entropy principle.

For a given entropy with fixed state variables $\boldsymbol{\epsilon}$, of all the values of the variable E, a state of equilibrium corresponds to the minimum energy. This is known as the minimum energy principle.

The Gibbs equation can be written as

$$dE = TdS + \tau_i d\epsilon_i, \quad i = 1, 2, \ldots, N, \tag{9.42}$$

with

$$T = \frac{\partial E}{\partial S}, \quad \tau_i = \frac{\partial E}{\partial \epsilon_i}. \tag{9.43}$$

We define a thermodynamic path in the state space by using the parameter λ as

$$S = S(\lambda), \quad \epsilon_i = \epsilon_i(\lambda). \tag{9.44}$$

If S is independent of λ, the process is called isentropic. For an isothermal process, $T = \partial E/\partial S$ has to be independent of λ.

9.4.1 Legendre or Contact Transformation

There are situations for which it is convenient to control the temperature T instead of the entropy S or tensions $\tau_i, i = 1, 2, \ldots, M \le N$, instead of ϵ_i. We obtain new functions with T and τ_i as independent variables by using Legendre transformations or contact transformations.

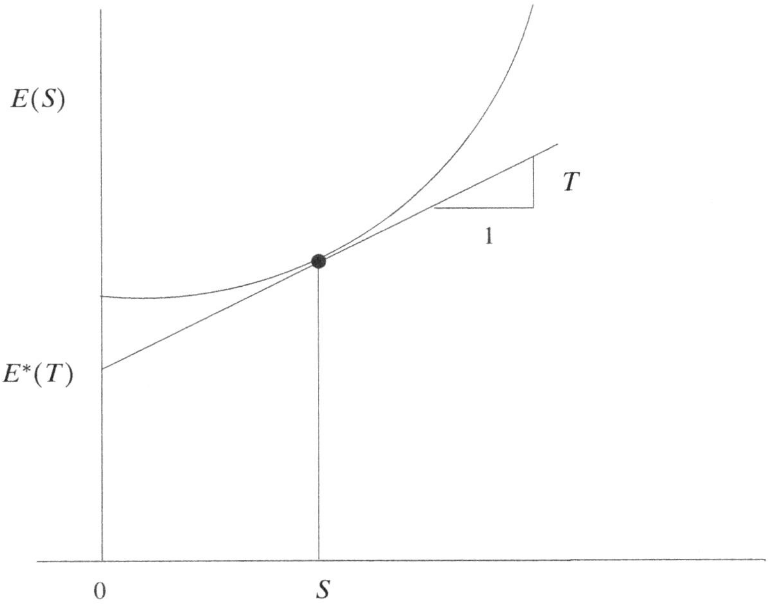

Figure 9.1. Legendre transform of $E(S)$ to $E^*(T)$.

As shown in Fig. 9.1, a plot of E as a function of S has to have a positive slope T, the absolute temperature. If the slope is given, we can move a line with that prescribed slope to contact the curve, E vs. S. If the intercept of this line with the E axis is denoted by $E^*(T)$, we have

$$E(S) = E^*(T) + TS, \quad E^*(T) = E(S) - TS. \tag{9.45}$$

From this, we obtain

$$\frac{\partial E^*}{\partial T} = -S, \quad \frac{\partial E}{\partial S} = T. \tag{9.46}$$

This process of introducing T as the independent variable is called a Legendre or contact transformation of $E(S)$. The term "contact" refers to the construction of $E^*(T)$ by contacting the E–S curve by a straight line of prescribed slope. In addition, if we want to use τ_i, $i = 1, 2, \ldots, M$ as new variables, we introduce

$$E^*(T, \tau_1, \tau_2, \ldots, \tau_M) = E(S, \epsilon_1, \epsilon_2, \ldots, \epsilon_N) - TS - \sum_{i=1}^{i=M} \tau_i \epsilon_i. \tag{9.47}$$

The following list shows some of the well-known functions in thermodynamics obtained by the Legendre transform. Here the summation on i from 1 to N is implied.

Potential	*Relation*	*Variables*	
Internal energy	E	S, ϵ_i	
Helmholtz free energy	$U = E - TS$	T, ϵ_i	
Enthalpy	$H = E - \tau_i \epsilon_i$	S, τ_i	
Gibbs free energy	$G = E - TS - \tau_i \epsilon_i$	T, τ_i	(9.48)

The following differential relations can be obtained from our list:

$$
\begin{aligned}
dE &= TdS + \tau_i d\epsilon_i, \\
dU &= -SdT + \tau_i d\epsilon_i, \\
dH &= TdS - \epsilon_i d\tau_i, \\
dG &= -SdT - \epsilon_i d\tau_i.
\end{aligned} \tag{9.49}
$$

We also use the notation for specific (per unit mass) quantities, ε for internal energy, u for Helmholtz free energy, h for enthalpy, and g for Gibbs free energy.

The Helmholtz free energy U is convenient to describe an isothermal process as the temperature T is an independent variable. Similarly, in solid mechanics for constant-stress processes, H and G are useful functions.

9.4.2 Surface Energy

In certain areas of continuum mechanics, the basic assumption of continuity of field variables has to be relaxed to include the creation of new surfaces of discontinuity inside the body. These surfaces are assumed to be formed at the expense of energy, and this energy is taken to be stored in the newly formed surface $\mathcal{S}_c$. The conservation of energy has to be written, including the surface energy, as

$$
\frac{D}{Dt}(K + E + \int_{\mathcal{S}_c} \gamma da) = P + H, \tag{9.50}
$$

where γ is the surface energy density. There is also an entropy production associated with the creation of new surfaces. To our previous entropy production rates we add a surface entropy term to get

$$
\frac{DS}{Dt} = \frac{D}{Dt}(S_E + S_I + \int_{\mathcal{S}_c} \eta da), \tag{9.51}
$$

where η is the surface entropy density.

9.5 Method of Jacobians in Thermodynamics

A method of manipulating thermodynamic functions by use of Jacobians was presented by Shaw in 1935 (see Margenau and Murphy, 1943). In the derivation of many thermodynamic relations, independent variables are changed frequently from one set to another. The Jacobian determinant of such transformations can be effectively used as an operational tool to achieve the desired result in a more direct fashion.

Consider a mapping from a two-dimensional (2D) space (x, y) to another space (u, v). Assuming a one-to-one mapping, we have

$$
dudv = Jdxdy, \tag{9.52}
$$

where J is the Jacobian determinant:

$$J = \begin{vmatrix} \left.\frac{\partial u}{\partial x}\right|_y & \left.\frac{\partial u}{\partial y}\right|_x \\ \left.\frac{\partial v}{\partial x}\right|_y & \left.\frac{\partial v}{\partial y}\right|_x \end{vmatrix}. \tag{9.53}$$

We use a shorthand notation:

$$J = \frac{[u, v]}{[x, y]}. \tag{9.54}$$

Although the individual bracket group in the numerator has no meaning without the denominator, symbolically we have

$$[u, v] = J[x, y], \tag{9.55}$$

and, based on the determinant,

$$[u, v] = -[v, u], \quad [u, u] = 0, \tag{9.56}$$

and as a product of two determinants,

$$\frac{[u, v]}{[x, y]}\frac{[x, y]}{[p, q]} = \frac{[u, v]}{[p, q]}, \tag{9.57}$$

where we have used two transformations

$$(u, v) \Rightarrow (x, y) \Rightarrow (p, q). \tag{9.58}$$

In this notation, the partial derivative

$$\left.\frac{\partial u}{\partial x}\right|_y = \frac{[u, y]}{[x, y]}. \tag{9.59}$$

A cyclic property of this notation can be obtained as follows: Let $w = w(u, v)$. Then

$$\begin{aligned} dw &= \left.\frac{\partial w}{\partial u}\right|_v du + \left.\frac{\partial w}{\partial v}\right|_u dv \\ &= \frac{[w, v]}{[u, v]}du + \frac{[w, u]}{[v, u]}dv. \end{aligned} \tag{9.60}$$

Further, assume that $u = u(x, y)$ and $v = v(x, y)$:

$$\begin{aligned} \left.\frac{\partial w}{\partial x}\right|_y &= \frac{[w, v]}{[u, v]}\left.\frac{\partial u}{\partial x}\right|_y + \frac{[w, u]}{[v, u]}\left.\frac{\partial v}{\partial x}\right|_y, \\ \frac{[w, y]}{[x, y]} &= \frac{[w, v]}{[u, v]}\frac{[u, y]}{[x, y]} + \frac{[w, u]}{[v, u]}\frac{[v, y]}{[x, y]}. \end{aligned} \tag{9.61}$$

After canceling $[x, y]$, and changing y to x, we have the cyclic relation

$$[u, v][w, x] + [v, w][u, x] + [w, u][v, x] = 0. \tag{9.62}$$

Having familiarized ourselves with this notation, let us apply it to thermodynamic variables.

To begin with, we have the Gibbs relation

$$d\varepsilon = Tds - pdv, \quad T = \left.\frac{\partial \varepsilon}{\partial s}\right|_v, \quad -p = \left.\frac{\partial \varepsilon}{\partial v}\right|_s, \tag{9.63}$$

where v is the specific volume. Using the equality of mixed derivatives, we obtain the Maxwell relation

$$\begin{aligned} \left.\frac{\partial T}{\partial v}\right|_s &= \left.\frac{\partial(-p)}{\partial s}\right|_v, \\ \frac{[T,s]}{[v,s]} &= -\frac{[p,v]}{[s,v]} = \frac{[p,v]}{[v,s]}, \\ [T,s] &= [p,v]. \end{aligned} \tag{9.64}$$

This shows that, if we map from the T, s plane to the p, v plane, the area elements remain constant.

Now we may derive further Maxwell relations:

$$\begin{aligned} \left.\frac{\partial p}{\partial T}\right|_s &= \frac{[p,s]}{[T,s]} = \frac{[p,s]}{[p,v]} = \left.\frac{\partial s}{\partial v}\right|_p, \\ \left.\frac{\partial p}{\partial s}\right|_T &= \frac{[p,T]}{[s,T]} = -\frac{[p,T]}{[p,v]} = -\left.\frac{\partial T}{\partial v}\right|_p, \\ \left.\frac{\partial v}{\partial T}\right|_s &= \frac{[v,s]}{[T,s]} = \frac{[v,s]}{[p,v]} = -\left.\frac{\partial s}{\partial p}\right|_v, \\ \left.\frac{\partial v}{\partial s}\right|_T &= \frac{[v,T]}{[s,T]} = \frac{[v,T]}{[v,p]} = \left.\frac{\partial T}{\partial p}\right|_v. \end{aligned} \tag{9.65}$$

For an arbitrary fixed variable x, we have

$$\begin{aligned} [\varepsilon, x] &= T[s,x] - p[v,x], \\ [h,x] &= T[s,x] + v[p,x]. \end{aligned} \tag{9.66}$$

The specific heats at constant pressure c_p and at constant volume c_v can be expressed as

$$\begin{aligned} c_p &= \left.\frac{\partial h}{\partial T}\right|_p = \frac{[h,p]}{[T,p]} = T\frac{[s,p]}{[T,p]} = T\left.\frac{\partial s}{\partial T}\right|_p, \\ c_v &= \left.\frac{\partial \varepsilon}{\partial T}\right|_v = \frac{[\varepsilon,v]}{[T,v]} = T\left.\frac{\partial s}{\partial T}\right|_v. \end{aligned} \tag{9.67}$$

From these,

$$\begin{aligned} c_p - c_v &= T\left[\frac{[s,p]}{[T,p]} - \frac{[s,v]}{[T,v]}\right] \\ &= T\frac{[s,p][T,v] - [s,v][T,p]}{[T,p][T,v]}. \end{aligned} \tag{9.68}$$

Using our cyclic identity

$$[p,v][T,s] + [v,T][p,s] + [T,p][v,s] = 0, \tag{9.69}$$

we find

$$c_p - c_v = -T\frac{[p, v]}{[T, p]}\frac{[p, v]}{[T, v]} = T\frac{\partial v}{\partial T}\bigg|_p\frac{\partial p}{\partial T}\bigg|_v. \tag{9.70}$$

Similarly,

$$\frac{c_p}{c_v} = \frac{[s, p]}{[s, v]}\frac{[T, v]}{[T, p]} = \frac{\partial p}{\partial v}\bigg|_s\frac{\partial v}{\partial p}\bigg|_T. \tag{9.71}$$

For an ideal gas,

$$pv = RT, \tag{9.72}$$

and we have

$$c_p - c_v = R. \tag{9.73}$$

SUGGESTED READING

Çengel, Y. A. and Boles, M. (2002). *Thermodynamics*, 4th ed., McGraw-Hill.

Christoffersen, J., Mehrabadi, M. M., and Nemat-Nasser, S. (1981). A micromechanical description of granular material behavior, *J. Appl. Mech.*, **48**, 339–344.

Cowin, S. C. (1978). Microstructural continuum models for granular materials, in *Proceedings of the U.S. Japan Seminar on Continuum Mechanical and Statistical Approaches in the Mechanics of Granular Materials*, Gakujatsu Bunken Fukyukai, pp. 162–170.

Margenau, H. and Murphy, G. M. (1943). *The Mathematics of Physics and Chemistry*, Van Nostrand.

Sonntag, R. E., Borgnakke, C., and Van Wylen, G. (2003). *Fundamentals of Thermodynamics*, 6th ed., Wiley.

EXERCISES

9.1. For an ideal gas,

$$\varepsilon = CT, \quad pv = RT.$$

Obtain the functions $s(T, v)$, $s(T, p)$, and $\varepsilon(s, v)$ for this substance, explicitly.

9.2. The internal energy and specific volume are given as

$$\varepsilon = \varepsilon(T, v), \quad v = v(T, p).$$

(a) Derive expressions for specific heat at constant pressure c_p and constant volume c_v, and show that

$$c_p - c_v = \frac{\partial v}{\partial T}\bigg|_p\left[p + \frac{\partial \varepsilon}{\partial v}\bigg|_T\right].$$

(b) Evaluate $c_p - c_v$ for the ideal gas.

(c) In the case of an adiabatic process ($\delta Q = 0$), show that an ideal gas satisfies

$$(c_v + R)pdv + c_v v dp = 0.$$

(d) From the preceding part, obtain pv^γ = constant, where $\gamma = c_p/c_v$.

9.3. In linear, isotropic thermoelasticity the Helmholtz free energy is given by

$$\rho u = G\epsilon_{ij}\epsilon_{ij} + \frac{\lambda}{2}\epsilon_{ii}\epsilon_{jj} - 2G\frac{1+\nu}{1-2\nu}\alpha(T - T_0)\epsilon_{ii},$$

where ρ, G, λ, ν, and α are constants. Assuming that ϵ_{ij}, the infinitesimal strain components, are the thermodynamic state variables, obtain the corresponding stresses, σ_{ij}.

9.4. In a Newtonian fluid, the hyperelastic part of the stress σ_{ij}^0 and the irreversible part σ_{ij}^{I} are given by

$$\sigma_{ij}^0 = -p\delta_{ij}, \quad \sigma_{ij}^{\mathrm{I}} = \lambda d_{kk}\delta_{ij} + 2\mu d_{ij},$$

where λ and μ are constants. Obtain the specific entropy production rate for this flow at constant temperature T.

9.5. If we assume the strain energy density for a material in the current configuration (in the absence of couple stresses) is given as

$$u = u(\partial_1 x_1, \partial_2 x_1, \ldots, \partial_3 x_3),$$

show that the first Piola–Kirchhoff stress tensor is obtained as

$$T_{\alpha i}^0 = \rho_0 \frac{\partial u}{\partial x_{i,\alpha}},$$

where we have used $x_{i,\alpha} = \partial x_i/\partial X_\alpha$.

9.6. Consider two neighboring equilibrium states I and II of a system. In state I, it has the state variables E, τ_i^0, ϵ_i, S, and T, and in state II, $E + \Delta E$, $\tau_i^0 + \Delta\tau_i^0$, $\epsilon_i + \Delta\epsilon_i$, $S + \Delta S$, and $T + \Delta T$.

(a) It is further assumed that no work is done on the system and no heat is added. If we add an assumption to the second law that what is permitted by the second law would actually occur in nature, show that the entropy at stable equilibrium state is a maximum.

(b) If we assume E is a monotonically increasing function of S, and between states I and II there is no change in entropy and $\Delta E \neq 0$, show that the stable equilibrium requires that E be a minimum.

9.7. An ideal rubber is considered to be incompressible, and its specific entropy change from a reference value of s_0 is given as

$$\rho(s - s_0) = -\kappa(\lambda_1^2 + \lambda_2^2 + \lambda_3^2 - 3),$$

where ρ is the constant density and λ_i is the stretch in the x_i direction. It is also known that at constant temperature, $\partial\varepsilon/\partial\lambda_i = 0$, where ε is the internal energy density. Obtain a relation between the second Piola–Kirchhoff stress S_{11} in terms of λ_1 for a band made of this material. The stress and stretch λ_1 are along the length of the band.

9.8. For a one-dimensional bar the Gibbs function is given as

$$G = G(T, \sigma), \quad s = -G_T, \quad \epsilon = -G_\sigma,$$

where the subscripts indicate partial differentiation, T is the absolute temperature, σ is the tensile stress, and ϵ is the infinitesimal strain along the length of the bar. If the bar has an initial length L and cross-sectional area of unity, show that the heat added δQ at constant temperature for a stress increase $\delta\sigma$, change in length δL, and temperature rise δT at constant stress are related in the form

$$\left.\frac{\delta Q}{\delta\sigma}\right|_T = \left.\frac{T}{L}\frac{\delta L}{\delta T}\right|_\sigma .$$

9.9. Show that

(a)

$$[s, \varepsilon] = -p[s, v], \quad [h, s] = v[p, s].$$

(b)

$$\left.\frac{\partial\varepsilon}{\partial p}\right|_s = -\frac{p}{\gamma}\left.\frac{\partial v}{\partial p}\right|_T, \quad \left.\frac{\partial\varepsilon}{\partial T}\right|_s = \frac{pc_v}{T}\left.\frac{\partial T}{\partial p}\right|_v,$$

where $\gamma = c_p/c_v$.

(c)

$$\left.\frac{\partial h}{\partial v}\right|_s = \gamma v\left.\frac{\partial p}{\partial v}\right|_T, \quad \left.\frac{\partial h}{\partial T}\right|_s = \frac{vc_p}{T}\left.\frac{\partial T}{\partial v}\right|_p .$$

(d)

$$\left.\frac{\partial T}{\partial s}\right|_\varepsilon = \frac{T}{c_v}\left[1 - \frac{T}{p}\left.\frac{\partial p}{\partial T}\right|_v\right], \quad \left.\frac{\partial T}{\partial s}\right|_h = \frac{T}{c_p}\left[1 - \frac{T}{v}\left.\frac{\partial v}{\partial T}\right|_p\right].$$

(e)

$$\left.\frac{\partial c_p}{\partial p}\right|_T = -T\left.\frac{\partial^2 v}{\partial T^2}\right|_p, \quad \left.\frac{\partial c_v}{\partial v}\right|_T = T\left.\frac{\partial^2 p}{\partial T^2}\right|_v .$$

10 Constitutive Relations

Equations describing the response of specific materials to applied loads are called constitutive relations. So far, we have studied the local descriptions of deformation, motion, forces, and energy and entropy constraints. Subject to the same external forces, different materials will respond differently even if they have the same mass and geometric properties. This complex material behavior cannot be expressed by a single universal equation. Excluding chemical and electrical variables, we have the following conservation and balance relations:

$$\frac{\partial \rho}{\partial t} + \partial_i(\rho v_i) = 0,$$

$$\partial_i \sigma_{ij} + \rho(f_j - a_j) = 0,$$

$$\partial_i m_{ij} + e_{jkl}\sigma_{kl} + \rho(l_j - \alpha_j) = 0,$$

$$\rho\dot{\varepsilon} - \sigma_{ij}\partial_i v_j - m_{ij}\partial_i \mu_j - \partial_i q_i - \rho h = 0, \tag{10.1}$$

which form eight equations. We have ρ, v_i, μ_i, σ_{ij}, m_{ij}, q_i, ε, and, implicitly, the temperature field T as unknowns. These add up to 30 unknowns. The accelerations a_i and α_i are derived from v_i and μ_i through the kinematic relations discussed in Chapter 8. Of the 22 additional relations required for completing the problem, an equation of state for ε is one; we need 9 relations for σ_{ij} and 9 for m_{ij}, which leaves 3. These are the relations for the flux q_i. These 22 relations are called the constitutive relations.

One approach to obtain these needed relations is to use the known laws relating forces and displacement of molecules in a lattice structure or in a multiply colliding fluid situation. Progress in this direction is being made by use of molecular dynamics and the Monte Carlo simulations. The number of particles one could introduce in these simulations is still moderate because of computational limitations.

Another approach is through experiments. However, the number of experiments required is still prohibitively large. The preferred method is known as phenomenology. In the phenomenological approach, mathematical expressions relating stress and deformation variables are proposed subject to certain invariance principles required by continuum mechanics. These relations are tested and refined by a limited number of experiments. They are further generalized to include observed deviations in new materials.

10.1 Invariance Principles

The invariance principles may be grouped into six categories:

1. principles of exclusion,
2. principle of coordinate invariance,
3. principle of spatial invariance,
4. principle of material invariance,
5. principle of dimensional invariance, and
6. principle of consistency.

We examine briefly each of these items.

10.1.1 Principles of Exclusion

These principles provide, on the basis of experience, certain rules for excluding a large number of variables from constitutive relations. We may consider the following classes of exclusion principles under this heading.

1. **Principle of heredity**: The behavior of a material particle $\boldsymbol{X}$ at time t is dependent on only the past experience of the body. For example, for the Cauchy stress,

$$\boldsymbol{\sigma} = \mathop{\mathcal{F}}_{\tau=-\infty}^{\tau=t} [x(X, \tau)], \tag{10.2}$$

where $\mathcal{F}$ represents a general functional that may involve integrals in time. The basic idea is to exclude the future experience of the particle. As an example, consider a one-dimensional bar with the stress–strain relation

$$\sigma(t) = E\epsilon(t) + \alpha \int_{-\infty}^{t} e^{-\beta(t-\tau)} \frac{d\epsilon}{d\tau} d\tau. \tag{10.3}$$

Here we have a kernel inside the integral that shows the material has a fading memory of the past strain rate.

2. **Principle of neighborhood**: The behavior of a particle, $\boldsymbol{X}$, occupying a point, $\boldsymbol{x}$, in the current configuration at time, t, is only affected by the history of an arbitrarily small neighborhood of $\boldsymbol{x}$.

$$\begin{aligned}\boldsymbol{\sigma} &= \mathop{\mathcal{F}}_{\tau=-\infty}^{\tau=t} [dx(X, \tau)], \\ &= \mathop{\mathcal{F}}_{\tau=-\infty}^{\tau=t} [\partial_\alpha x(X, \tau), \partial_\alpha \partial_\beta x(X, \tau), \ldots,]. \end{aligned} \tag{10.4}$$

These two principles are called the principles of determinism. Modern theories that incorporate the influence of particles at a distance (nonlocal theories) have been found useful in certain applications. The additional principles needed for electromagnetic phenomena are not included here.

10.1.2 Principle of Coordinate Invariance

This simply restricts the variables to the tensor format, which automatically satisfies coordinate invariance. For example, if we use the system x, we have

$$\sigma_{ij} = \mathop{\mathcal{F}_{ij}}_{\tau=-\infty}^{\tau=t} [\partial_\alpha x_k(X, \tau)]. \tag{10.5}$$

In a new system x',

$$\sigma'_{ij} = \mathop{\mathcal{F}'_{ij}}_{\tau=-\infty}^{\tau=t} [\partial_\alpha x'_k(X, \tau)]. \tag{10.6}$$

10.1.3 Principle of Spatial Invariance

The constitutive relations must be invariant under rigid body translations and rotations of the spatial coordinates. In other words, the tensors entering these equations must be objective tensors. For example, consider the case of a simple material for which $dx \Rightarrow \partial_\alpha x_k$. We may use coordinate system x' or x to describe the constitutive relations. As the material responds, irrespective of the spatial system we use, first we have

$$\mathcal{F}_{ij} = \mathcal{F}'_{ij}. \tag{10.7}$$

As far as the independent variables are concerned, these two systems are related as

$$x'_i = Q_{ij} x_j, \quad x_i = Q_{ji} x'_j. \tag{10.8}$$

The relation

$$\sigma_{ij} = \mathop{\mathcal{F}_{ij}}_{\tau=-\infty}^{\tau=t} [\partial_\alpha x_k, \tau] \tag{10.9}$$

transforms to

$$\sigma'_{ij} = \mathop{\mathcal{F}_{ij}}_{\tau=-\infty}^{\tau=t} [\partial_\alpha x'_k(X, \tau)],$$

$$Q_{im} Q_{jn} \sigma_{mn} = \mathop{\mathcal{F}_{ij}}_{\tau=-\infty}^{\tau=t} [\partial_\alpha Q_{kl} x_l(X, \tau)],$$

$$\sigma_{mn} = Q_{im} Q_{jn} \mathop{\mathcal{F}_{ij}}_{\tau=-\infty}^{\tau=t} [\partial_\alpha Q_{kl} x_l(X, \tau)]. \tag{10.10}$$

Comparing this with Eq. (10.9), we find a constraint on the allowable functional forms of $\mathcal{F}_{ij}$,

$$\mathop{\mathcal{F}_{mn}}_{\tau=-\infty}^{\tau=t} [\partial_\alpha x_k(X, \tau)] = Q_{im} Q_{jn} \mathop{\mathcal{F}_{ij}}_{\tau=-\infty}^{\tau=t} [Q_{kl} \partial_\alpha x_l(X, \tau)]. \tag{10.11}$$

In tensor notation, we have (omitting the hereditary integral)

$$\mathcal{F}[\nabla \boldsymbol{x}] = \boldsymbol{Q}^T \cdot \mathcal{F}[\nabla \boldsymbol{x} \cdot \boldsymbol{Q}^T] \cdot \boldsymbol{Q}. \tag{10.12}$$

All variables have to be objective.

10.1.4 Principle of Material Invariance

We may transform the material coordinates X_α by using proper orthogonal matrices $\boldsymbol{Q}$. This may be generalized to matrices $\boldsymbol{R}$ (not to be confused with $\boldsymbol{R}$ in polar decomposition), called the full orthogonal group of transformations (see Ludwig and Falter, 1987) if they impart not only rotations but also reflections of coordinates:

$$X'_\alpha = R_{\alpha\beta} X_\beta. \tag{10.13}$$

If the material has certain symmetry, such as isotropy, orthotropy, etc., the constitutive relations must be invariant with respect to transformations that obey this symmetry. For example, isotropic constitutive relations must be invariant with respect to the full orthogonal group. Transversely isotropic materials allow rotation about a fixed axis. There are crystals with hexagonal packing; these should have constitutive relations that are invariant if the material coordinates are rotated by 60°.

10.1.5 Principle of Dimensional Invariance

In their functional forms, constitutive relations must obey dimensional constraints.

10.1.6 Principle of Consistency

The basic principles of conservation and balance laws, as well as the entropy production inequality, must not be violated by constitutive equations.

The importance of the preceding principles can be appreciated if we visualize an experimentalist recording the response of one variable, say the length of a bar, to the applied load. We readily see a plot of the current length versus the applied load. If we want to generalize this relation to bars of different cross-sectional areas and lengths, the concepts of stress and strain enter the scheme. Next, we want to generalize this to three dimensions by incorporating anisotropy and nonlinear stress–strain relations, subject to the principles just discussed. The situation is more complicated for fluids and viscoelastic materials.

10.2 Simple Materials

This group of materials is not as simple as the name implies! In the principle of neighborhood, we include only the deformation gradient; all other higher derivatives are neglected. A large class of materials for which the stresses depend on the history of the deformation gradient $\boldsymbol{F}$ is included here. We begin with

$$\sigma = \mathop{\mathcal{F}}_{\tau=-\infty}^{\tau=t} [\partial_\alpha x_k]. \tag{10.14}$$

This relation, when subjected to the principle of spatial invariance, gives

$$\boldsymbol{Q} \cdot \boldsymbol{\sigma} \cdot \boldsymbol{Q}^T = \mathop{\mathcal{F}}_{\tau=-\infty}^{\tau=t} [\boldsymbol{F} \cdot \boldsymbol{Q}^T]. \tag{10.15}$$

Next, we use the polar decomposition,

$$\boldsymbol{F} = \boldsymbol{V} \cdot \boldsymbol{R}, \tag{10.16}$$

to write this relation as

$$\boldsymbol{Q} \cdot \boldsymbol{\sigma} \cdot \boldsymbol{Q}^T = \mathop{\mathcal{F}}_{\tau=-\infty}^{\tau=t} [\boldsymbol{V} \cdot \boldsymbol{R} \cdot \boldsymbol{Q}^T]. \tag{10.17}$$

The orthogonal, time-dependent tensor $\boldsymbol{R}$ is arbitrary. We may choose

$$\boldsymbol{Q} = \boldsymbol{R} \tag{10.18}$$

to simplify the preceding relation to obtain

$$\boldsymbol{R} \cdot \boldsymbol{\sigma} \cdot \boldsymbol{R}^T = \mathop{\mathcal{F}}_{\tau=-\infty}^{\tau=t} [\boldsymbol{V}] \tag{10.19}$$

or

$$\boldsymbol{\sigma} = \boldsymbol{R}^T \cdot \mathop{\mathcal{F}}_{\tau=-\infty}^{\tau=t} [\boldsymbol{V}] \cdot \boldsymbol{R}. \tag{10.20}$$

This shows that the Cauchy stress depends on the history of the left-stretch tensor $\boldsymbol{V}$ only and not on the history of the rotation $\boldsymbol{R}$.

As

$$\boldsymbol{V} = \boldsymbol{G}^{1/2}, \tag{10.21}$$

the preceding functional form can also be written as

$$\boldsymbol{\sigma} = \boldsymbol{R}^T \cdot \mathop{\mathcal{G}}_{\tau=-\infty}^{\tau=t} [\boldsymbol{G}] \cdot \boldsymbol{R}. \tag{10.22}$$

A further change in the functional form can be obtained with the Lagrange strain $\boldsymbol{E}$, defined by

$$\boldsymbol{G} = 2\boldsymbol{E} + \boldsymbol{I}. \tag{10.23}$$

Instead of the general history-dependent **functionals**, as limiting cases, we have $\mathcal{G}$ as simple functions. For example,

1. elastic materials: $\mathcal{G}$ is a function of $\boldsymbol{E}$ only;
2. viscoelastic materials: $\mathcal{G}$ is a function of $\boldsymbol{E}$ and $\dot{\boldsymbol{E}}$;
3. Stokes fluids: $\mathcal{G}$ is a function of $\dot{\boldsymbol{E}}$.

10.3 Elastic Materials

There are two ways of considering elastic materials: One is by using Cauchy's definition and the other is by using the Green's definition. The couple stresses are omitted for this discussion.

10.3.1 Elastic Materials of Cauchy

Special case 1, in the preceding list, can be written as

$$\sigma = \mathcal{G}(\partial_\alpha x_k). \tag{10.24}$$

We stipulate that there exists a natural state of the material that is distinguished by uniform conditions inside the body. In the natural state the deformation gradient is zero and the body is unstressed. Recent theories consider natural states evolving with the deformation history. We consider a unique natural state. This implies that at time $t = 0$, the Green's deformation tensor is the identity tensor, i.e.,

$$\boldsymbol{G}(0) = \boldsymbol{I}. \tag{10.25}$$

As the couple stress tensor is zero, $\boldsymbol{\sigma}$ is symmetric and so is $\mathcal{G}$. From our study of spatial invariance, we can write

$$\boldsymbol{\sigma} = \boldsymbol{R}^T \cdot \mathcal{F}(\boldsymbol{G}) \cdot \boldsymbol{R}, \tag{10.26}$$

where $\mathcal{G}(\boldsymbol{G}^{1/2}) = \mathcal{F}(\boldsymbol{G})$, and $\boldsymbol{R}$ is the rotation tensor in the polar decomposition of the deformation gradient tensor $\boldsymbol{F}$.

For anisotropic materials, the orientation of the undeformed material has to be distinguished, and the function $\mathcal{F}$ would depend on the initial orientation. As we have seen, the initial orientation of the unit vector $\boldsymbol{e}_\alpha$ is now $\boldsymbol{G}_\alpha$ in the current configuration. We write this in the form

$$\mathcal{F} = \mathcal{F}(\boldsymbol{G}, \boldsymbol{G}_\alpha). \tag{10.27}$$

A third generalization is to make the material inhomogeneous. Now properties depend on the particle:

$$\mathcal{F} = \mathcal{F}(\boldsymbol{G}, \boldsymbol{G}_\alpha, X). \tag{10.28}$$

The arbitrary matrix function $\mathcal{F}$, with the 3×3 matrix $\boldsymbol{G}$ as its argument, can be represented by the Cayley–Hamilton theorem as

$$\mathcal{F} = c_0'\boldsymbol{I} + c_1'\boldsymbol{G} + c_2'\boldsymbol{G}^2, \tag{10.29}$$

where the coefficients c_i' are functions of the eigenvalues of $\boldsymbol{G}$.

We may multiply Eq. (10.26) by $\boldsymbol{f} = \boldsymbol{F}^{-1} = \boldsymbol{V}^{-1}\boldsymbol{R}^T$ to obtain the form

$$\boldsymbol{S} = \sqrt{I_{G3}}\boldsymbol{f}^T \cdot \boldsymbol{\sigma} \cdot \boldsymbol{f} = c_0\boldsymbol{G}^{-1} + c_1\boldsymbol{I} + c_2\boldsymbol{G}, \tag{10.30}$$

where $c_i = \sqrt{I_{G3}}c_i'$ and $\boldsymbol{S}$ is the second Piola–Kirchhoff stress.

In finite-strain elasticity, an area of intense research activity has been rubber elasticity. Rubber is considered to be an incompressible material up to moderately high strains. When a material is incompressible, there is no volume change and $I_{G3} = 1$. But it can support an arbitrary amount of pressure, which is called a reactive stress as it cannot be found from the constitutive relations. Then the stress can be written as

$$\boldsymbol{S} = -\overline{p}\boldsymbol{I} + c_0\boldsymbol{G}^{-1} + c_1\boldsymbol{I} + c_2\boldsymbol{G}, \tag{10.31}$$

where, now, $c_i = c_i(I_{G1}, I_{G2})$.

10.3.2 Elastic Materials of Green

According to the Green's definition of elasticity, elastic materials have an internal energy function called the strain energy, and the stresses can be derived from it by using appropriate strain measures as state variables. This group of elastic materials are also known as **hyperelastic**. For an isothermal case in which the temperature T is fixed, the appropriate internal energy function is the Helmholtz free energy U. If ϵ_i are the state variables representing strain, the conjugate thermodynamic tensions are

$$\tau_i^0 = \rho \frac{\partial u}{\partial \epsilon_i}\bigg|_T, \tag{10.32}$$

where u is the Helmholtz free energy density. The scalar function u has to be invariant with respect to the rotation of the spatial coordinates. If we take $\partial_\alpha x_i$ as a measure of the strain ϵ_i and assume $u = u(\partial_\alpha x_i)$, we see that a rotation of the spatial coordinates would alter the form of u. Because $dx_i dx_i = (ds)^2$ is an invariant, we assume

$$u = u(\partial_\alpha x_i \partial_\beta x_i) = u(G_{\alpha\beta}). \tag{10.33}$$

This functional form for the Helmholtz free energy is known as the Boussinesq form. Then, with

$$\epsilon_i \Rightarrow \partial_\alpha x_i, \quad \dot{\epsilon}_i \Rightarrow \partial_j v_i \partial_\alpha x_j, \tag{10.34}$$

we have

$$\tau_{\alpha i}^0 = 2\rho \frac{\partial u}{\partial G_{\alpha\beta}} \partial_\beta x_i. \tag{10.35}$$

From Chapter 9 we have

$$\sigma_{ij}^0 \partial_i v_j = \sigma_{ij}^0 d_{ij} = \rho \frac{\partial u}{\partial \epsilon_i} \dot{\epsilon}_i. \tag{10.36}$$

Substituting for ϵ_i and $\dot{\epsilon}_i$, we have

$$\sigma_{ij}^0 \partial_i v_j = 2\rho \frac{\partial u}{\partial G_{\alpha\beta}} \partial_\beta x_i \partial_\alpha x_j \partial_j v_i. \tag{10.37}$$

The expression on the right-hand side is symmetric in α and β because of the symmetry of $G_{\alpha\beta}$. Equating the coefficients of the velocity gradients from both sides, we obtain an expression for $\boldsymbol{\sigma}^0$ as

$$\sigma_{ij}^0 = 2\rho \frac{\partial u}{\partial G_{\alpha\beta}} \partial_\beta x_i \partial_\alpha x_j. \tag{10.38}$$

In terms of the second Piola–Kirchhoff stress, we have

$$S_{\alpha\beta} = 2\rho_0 \frac{\partial u}{\partial G_{\alpha\beta}}. \tag{10.39}$$

Further, if we take u as a function of the Green's strain $\boldsymbol{E}$,

$$S_{\alpha\beta} = \rho_0 \frac{\partial u}{\partial E_{\alpha\beta}}. \tag{10.40}$$

For isotropic, homogeneous materials, rotations of the initial configuration should not affect the form of the Helmholtz free energy function. Then u should depend on $\boldsymbol{E}$ only through the three invariants of $\boldsymbol{E}$ or those of $\boldsymbol{G}$. The three invariants of $\boldsymbol{G}$ are

$$\begin{aligned} I_{G1} &= G_{\alpha\beta}\delta_{\alpha\beta}, \\ I_{G2} &= \frac{1}{2}[G_{\alpha\alpha}G_{\beta\beta} - G_{\alpha\beta}G_{\alpha\beta}], \\ I_{G3} &= \frac{1}{6}[e_{\alpha\beta\gamma}e_{\mu\nu\lambda}G_{\alpha\mu}G_{\beta\nu}G_{\gamma\lambda}]. \end{aligned} \tag{10.41}$$

The derivatives of these invariants with respect to $G_{\alpha\beta}$ are obtained as

$$\begin{aligned} \frac{\partial I_{G1}}{\partial G_{\alpha\beta}} &= \delta_{\alpha\beta}, \\ \frac{\partial I_{G2}}{\partial G_{\alpha\beta}} &= I_{G1}\delta_{\alpha\beta} - G_{\alpha\beta}, \\ \frac{\partial I_{G3}}{\partial G_{\alpha\beta}} &= I_{G3}G_{\alpha\beta}^{-1}. \end{aligned} \tag{10.42}$$

The second Piola–Kirchhoff stress becomes

$$S_{\alpha\beta} = 2\rho_0[u_{,1}\,\delta_{\alpha\beta} + u_{,2}\,(I_{G1}\delta_{\alpha\beta} - G_{\alpha\beta}) + u_{,3}\,I_{G3}G_{\alpha\beta}^{-1}], \tag{10.43}$$

where we have used the notation

$$u_{,i} = \frac{\partial u}{\partial I_{Gi}} \tag{10.44}$$

Finally, we rearrange the terms to get $\boldsymbol{S}$ in the form

$$\boldsymbol{S} = g_0\boldsymbol{G}^{-1} + g_1\boldsymbol{I} + g_2\boldsymbol{G}, \tag{10.45}$$

where the coefficients are defined as

$$g_0 = 2\rho_0 I_{G3}u_{,3}, \quad g_1 = 2\rho_0(u_{,1} + I_{G1}u_{,2}), \quad g_2 = -2\rho_0 u_{,2}. \tag{10.46}$$

In the natural state, $\boldsymbol{G} = \boldsymbol{I}$, and

$$I_{G1} = 3, \quad I_{G2} = 3, \quad I_{G3} = 1. \tag{10.47}$$

For a stress-free natural state, for a Green's elastic material,

$$(g_0 + g_1 + g_2)|_{I_{G1}=3, I_{G2}=3, I_{G3}=1} = 0 \quad \text{or} \quad u_{,1} + 2u_{,2} + u_{,3} = 0, \tag{10.48}$$

and for a Cauchy elastic material,

$$(c_0 + c_1 + c_2)|_{I_{G1}=3, I_{G2}=3, I_{G3}=1} = 0. \tag{10.49}$$

The coefficients g_i for the Green's elastic material are obtained as derivatives of a single function u, the Helmholtz free energy, and the coefficients c_i in the Cauchy elastic case are three independent functions. Using the mixed derivatives, in the sense of Maxwell relations, we may relate the derivatives of g_i.

For the incompressible case, $u_{,3} = 0$, as $I_{G3} = 1$. We add a hydrostatic pressure and write the stress as

$$\boldsymbol{S} = -\overline{p}\boldsymbol{I} + g_0 \boldsymbol{G}^{-1} + g_1 \boldsymbol{I} + g_2 \boldsymbol{G}. \tag{10.50}$$

10.4 Stokes Fluids

In Stokes fluids, the stress tensor is split into a reversible part that can be derived from a potential and an irreversible part:

$$\boldsymbol{\sigma} = \boldsymbol{\sigma}^0 + \boldsymbol{\sigma}^I. \tag{10.51}$$

As in an ideal gas,

$$\boldsymbol{\sigma}^0 = -p\boldsymbol{I}. \tag{10.52}$$

where p is called the **thermodynamic pressure**. The irreversible part is assumed to be a function of the deformation rate tensor $\boldsymbol{d}$:

$$\boldsymbol{\sigma}^I = \mathcal{F}(\boldsymbol{d}). \tag{10.53}$$

As the deformation rate tensor $\boldsymbol{d}$ is an objective tensor, this form satisfies the relevant invariance requirements. Next, we use the Cayley–Hamilton theorem to represent the function $\mathcal{F}(\boldsymbol{d})$ as a polynomial in $\boldsymbol{d}$:

$$\mathcal{F}(\boldsymbol{d}) = a_0 \boldsymbol{I} + a_1 \boldsymbol{d} + a_2 \boldsymbol{d}^2, \tag{10.54}$$

where $a_i = a_i(I_{d1}, I_{d2}, I_{d3})$, with I_{di} being the invariants of $\boldsymbol{d}$.

The stress $\boldsymbol{\sigma}$ can be written as

$$\boldsymbol{\sigma} = (a_0 - p)\boldsymbol{I} + a_1 \boldsymbol{d} + a_2 \boldsymbol{d}^2. \tag{10.55}$$

When there is no motion, all the invariants of $\boldsymbol{d}$ are equal to zero and the stress is the hydrostatic stress that is due to the thermodynamic pressure. That is,

$$a_i(0, 0, 0) = 0. \tag{10.56}$$

For an **incompressible fluid**, $I_{d1} = 0$, and the pressure is a reactive quantity and it remains as a basic unknown. For this case, the stress can be written as

$$\boldsymbol{\sigma} = -\overline{p}\boldsymbol{I} + a_1\boldsymbol{d} + a_2\boldsymbol{d}^2, \tag{10.57}$$

where $\overline{p}$ is called the mechanical pressure, and

$$a_i = a_i(I_{d2}, I_{d3}). \tag{10.58}$$

To satisfy the Clausius–Duhem inequality about the entropy production,

$$\rho T\dot{s} = \sigma_{ij}^{I} d_{ij} > 0. \tag{10.59}$$

Using the constitutive relation, this inequality becomes

$$a_1 d_{ii} + a_2 d_{ij} d_{ji} + a_3 d_{ij} d_{jk} d_{ki} > 0, \tag{10.60}$$

where the three terms contain the first invariants of $\boldsymbol{d}$, $\boldsymbol{d}^2$, and $\boldsymbol{d}^3$, respectively. We can show that

$$I_{d^21} = I_{d1}^2 - 2I_{d2}, \quad I_{d^31} = I_{d1}^3 - 3I_{d1}I_{d2} + 3I_{d3}. \tag{10.61}$$

The entropy production constraint can be written in terms of these invariants as

$$a_1 I_{d1} + a_2(I_{d1}^2 - 2I_{d2}) + a_3(I_{d1}^3 - 3I_{d1}I_{d2} + 3I_{d3}) > 0. \tag{10.62}$$

The mechanical pressure $\overline{p}$ can be expressed as

$$\bar{p} = -\frac{1}{3}t_{ii} = p - a_1 - \frac{1}{3}a_2 I_{d1} - \frac{1}{3}a_3(I_{d1}^2 - 2I_{d2}). \tag{10.63}$$

The mechanical pressure $\overline{p}$ is the pressure one measures as the average of three normal stresses, and the thermodynamic pressure, as we have seen, is related to the internal energy ε. In the static case they are equal. The coefficients a_i, which participate in the production of entropy, are called phenomenological coefficients. They are in general functions of the state variables: temperature and specific volume.

We continue the discussion of elastic materials and Stokes fluids in the next two chapters.

10.5 Invariant Surface Integrals

The balance and conservation laws discussed in this chapter lead to certain invariant surface integrals that have many applications, in particular in fracture mechanics and fluid dynamics. We make the following simplifications at the outset: The Cauchy stress tensor is symmetric, and the moment stresses and the intrinsic rotations are absent.

The local forms of balance and conservation are

$$\begin{aligned} \frac{\partial \rho}{\partial t} + \partial_i(\rho v_i) &= 0, \\ \partial_j \sigma_{ij} + \rho(f_i - a_i) &= 0, \\ \rho\dot{\varepsilon} - \sigma_{ij}\partial_j v_i - \partial_i q_i - \rho h &= 0. \end{aligned} \tag{10.64}$$

We anticipate applications for which, with respect to a steadily moving coordinate system, the field variables appear time independent. As an example, in front of an extending crack (far away from any boundaries) the stress field can be approximated as independent of time. An airfoil moving in a fluid will experience a time-independent velocity field around it. Let us introduce the moving coordinates through

$$x_i' = x_i - V_i t. \tag{10.65}$$

Then

$$\frac{\partial}{\partial t} = -V_i \frac{\partial}{\partial x_i'}, \quad \frac{D}{Dt} = (v_i - V_i)\frac{\partial}{\partial x_i'}. \tag{10.66}$$

Using the notation $\partial_i' = \partial/\partial x_i'$, conservation of mass, balance of momentum, and balance of energy can be written as

$$\begin{aligned} \partial_i'[\rho(v_i - V_i)] &= 0, \\ \partial_j'\sigma_{ij} - \rho(v_j - V_j)\partial_j'(v_i - V_i) + \rho f_i &= 0, \\ \rho(v_i - V_i)\partial_i'\varepsilon - \rho h - \partial_i' q_i - \sigma_{ij}\partial_j'(v_i - V_i) &= 0. \end{aligned} \tag{10.67}$$

Next, we integrate these three expressions over a fixed, arbitrary volume V' enclosed by a surface S' (in the moving coordinate system), with the assumption that none of the field quantities is singular inside this volume. We then convert the volume integrals into surface integrals by using the Gauss theorem. Thus the conservation of mass gives

$$\int_{S'} \rho(v_i - V_i)n_i' dS' = 0. \tag{10.68}$$

The momentum equation gives

$$\int_{S'} \{[\sigma_{ij} - \rho(v_j - V_j)(v_i - V_i)]n_j' - \phi n_i'\} \, dS' = 0, \tag{10.69}$$

where the potential ϕ is related to the body force through

$$\rho f_i = -\partial_i' \phi. \tag{10.70}$$

Multiplying the local form of the momentum equation by $(v_i - V_i)$, we find

$$(v_i - V_i)\partial_i'\sigma_{ij} - (v_i - V_i)\partial_i'\phi - \frac{1}{2}\rho(v_j - V_j)\partial_j'[(v_i - V_i)(v_i - V_i)] = 0. \tag{10.71}$$

Using the local forms of conservation of mass and the two forms of the momentum equations, we can write the energy equation as

$$\int_{S'} \left\{\rho(v_i - V_i)[\varepsilon + \frac{1}{2}(v^2 - V^2)] + \psi_i + \bar{\phi}_i + v_i\phi - q_i - \sigma_{ij}v_j\right\} n_i' dS' = 0, \tag{10.72}$$

where we have introduced two more potentials, $\bar{\phi}$ and $\boldsymbol{\psi}$, through the relations

$$\phi\partial_i' v_i = -\partial_j'\bar{\phi}_j, \quad \rho h = -\partial_i'\psi_i. \tag{10.73}$$

We may obtain a simpler version of the invariant integral involving the energy under the following assumptions: the heat generation, the heat flow, and the body forces are zero and the deformation gradient is small compared with unity. These are suitable for small deformations of solids:

$$v_i = \dot{u}_i = (v_j - V_j)\partial'_j u_i \quad \text{or} \quad |v_i| \ll |V_i|. \tag{10.74}$$

Neglecting terms of the order of v_i compared with those of the order of V_i, we find that the energy equation gives

$$\int_{S'} [\rho(\varepsilon + \frac{1}{2}V^2)n'_i - \sigma_{kj}\partial'_i u_j n'_k] dS' = 0, \tag{10.75}$$

where we have canceled a common factor, $(-V_i)$. Noting that the integral involving V^2 is zero, we find

$$\int_S [\rho\varepsilon n_i - \sigma_{kj}\partial_i u_j n_k] dS = 0. \tag{10.76}$$

In applications involving fluid mechanics, if we use the original coordinates, we get

$$\int_S v_i [\rho(\varepsilon + \frac{1}{2}v^2)n_i - \sigma_{ij} n_j] dS = 0. \tag{10.77}$$

10.6 Singularities

It was mentioned earlier that the surface integrals, Eqs. (10.76) and (10.77), vanish if there are no singularities inside the surface. When there is a point singularity, these integrals may not be zero. If we consider two surfaces S_1 and S_2 enclosing the singularity, the integrals over these surfaces will be the same as there is no singularity in the region between them. Thus we may shrink the surface to an arbitrarily small size to enclose the singularity. The integrals represent certain intrinsic properties of the singularity. One may compare this situation with Cauchy's residue theorem. By examining Eq. (10.77), we see the integral is the loss of energy from the small volume per unit time. This energy must equal the work done by the fluid on the singularity; in other words, the energy dissipates at the singularity. We will revisit this idea later when we consider elasticity and fluid dynamics. The preceding discussion is based on the theory presented by Cherepanov (1979).

SUGGESTED READING

Cherepanov, G. P. (1979). *Mechanics of Brittle Fracture*, McGraw-Hill.

Ludwig, W. and Falter, C. (1987). *Symmetries in Physics*, Springer-Verlag.

Malvern, L. E. (1969). *Introduction to the Mechanics of a Continuous Medium*, Prentice-Hall.

Truesdell, C. (1960). *Principles of Continuum Mechanics*, Field Research Laboratory, Socony Mobil Oil Co., Dallas, TX.

Truesdell, C. and Noll, W. (1965). The nonlinear field theories of mechanics, in *Encyclopedia of Physics* (S. Flügge, ed.), Springer-Verlag, Vol. 3/3.

Truesdell, C. and Toupin, R. A. (1960). The classical field theories, in *Encyclopedia of Physics*, (S. Flügge, ed.), Springer-Verlag, Vol. 3/1.

EXERCISES

10.1. Express the first invariants of the matrices $\boldsymbol{A}^2$ and $\boldsymbol{A}^3$ in terms of the three invariants of the matrix $\boldsymbol{A}$.

10.2. Obtain the Maxwell relations among the first-order derivatives of the coefficients g_i for the Green's elastic materials.

10.3. For hyperelastic materials, if we assume the Neumann–Kirchhoff form,

$$u = u(\partial_\alpha x_i),$$

obtain the corresponding stress.

10.4. Another form for u is given by the Hamel form,

$$u = u(\partial_i X_\alpha).$$

Obtain the stress for this spatial description.

10.5. Yet another form for u is the Murnaghan form,

$$u = u(e_{ij}).$$

What is the stress for this case?

10.6. Based on the constitutive relations of Cauchy elastic materials, hyperelastic materials, and the Stokes fluids, show that the principal axes of the second Piola–Kirchhoff stress $\boldsymbol{S}$ coincide with those of the Green's deformation tensor $\boldsymbol{G}$ and those of the Cauchy stress $\boldsymbol{\sigma}$ coincide with those of the deformation rate tensor $\boldsymbol{d}$.

10.7. In an isotropic elastic material, λ_i are the three principal stretches and the corresponding principal stresses are σ_i, respectively. If the stretches are ordered as

$$\lambda_1 < \lambda_2 < \lambda_3,$$

obtain the conditions necessary for a similar ordering of the principal stresses.

10.8. From experiments, one concludes that the second Piola–Kirchhoff stress can be written in terms of the deformation gradient components as

$$S_{\alpha\beta} = A_{\alpha\beta\gamma i}\,\partial_\gamma x_i + B_{\alpha\beta\gamma\delta}\,\partial_\gamma x_i\,\partial_\delta x_i.$$

Using the principles of invariance, evaluate the suitability of this relation.

10.9. In a shear flow the only nonzero deformation rate component is $\partial_2 v_1 = d$. The shear stress $\sigma_{21} = \tau$ at a point was measured for various values of d. Based on this study, it was suggested that

$$\tau = \mu\, d + \nu\, d^3,$$

where μ and ν are constants. How would you generalize this for the three-dimensional (3D) case for further experiments to verify your extrapolation?

11 Hyperelastic Materials

In the previous chapter we briefly examined the constitutive relations of hyperelastic materials. There has been a tremendous amount of research in the area of elastic materials, beginning with the early studies by Hooke, Navier, Cauchy, Germain, Kirchhoff, St. Venant, Airy, Love, and others. Most of these studies dealt with linear stress–strain relations with the infinitesimal-strain assumption. With the advent of the finite-element method and high-speed computers, the idealized geometrical configurations of the elastic bodies as well as the constraints of linear elasticity of the past can now be relaxed whenever needed.

We begin this chapter with a discussion of finite elasticity in a general setting. Some of the well-known inverse solutions are introduced. A small section is devoted to linear elasticity, as most students have had multiple courses in this area. The chapter concludes with the topic of linear thermoelasticity.

11.1 Finite Elasticity

The basic equations of finite elasticity, in terms of the Cauchy stress, can be listed as follows:

$$\begin{aligned}
&\text{conservation of mass:} && \rho = \rho_0/\sqrt{I_3}, \\
&\text{linear momentum:} && \nabla \cdot \boldsymbol{\sigma} = \rho(\boldsymbol{a} - \boldsymbol{f}), \\
&\text{angular momentum:} && \boldsymbol{\sigma} = \boldsymbol{\sigma}^T, \\
&\text{constitutive relation:} && \boldsymbol{\sigma} = -p\boldsymbol{I} + \boldsymbol{F}^T \cdot (a_0\boldsymbol{G}^{-1} + a_1\boldsymbol{I} + a_2\boldsymbol{G}) \cdot \boldsymbol{F},
\end{aligned} \tag{11.1}$$

where the coefficients a_i are given in terms of the partial derivatives of the strain energy density function (Helmholtz free energy density), $u(I_1, I_2, I_3)$, by

$$\begin{aligned}
a_0 &= 2\rho_0\sqrt{I_3}u_{,3}\,, \\
a_1 &= \frac{2\rho_0}{\sqrt{I_3}}(u_{,1} + I_1 u_{,2}\,), \\
a_2 &= -\frac{2\rho_0}{\sqrt{I_3}}u_{,2}\,.
\end{aligned} \tag{11.2}$$

We have written the constitutive relation in a form applicable to both compressible and incompressible materials. For compressible materials, $p = 0$, and for incompressible materials, $a_0 = 0$, and a_1 and a_2 are functions of I_1 and I_2 only, as $I_3 = 1$. We omit the subscript G for the invariants, as all invariants in this chapter pertain to the Green's deformation tensor. We also have the definitions

$$\boldsymbol{G} = \boldsymbol{F}\boldsymbol{F}^T, \quad F_{\alpha i} = \partial_\alpha x_i, \quad \boldsymbol{a} = \ddot{\boldsymbol{x}}. \tag{11.3}$$

At this stage, we keep the Helmholtz free energy density u as an unspecified function. For rubberlike materials, which are assumed to be incompressible, u is often assumed in the form

$$u = A_1(I_1 - 3) + A_2(I_2 - 3), \tag{11.4}$$

with A_i being constants. Materials with this Helmholtz energy function are known as Mooney–Rivlin materials. If $A_2 = 0$, we have what are known as Mooney materials.

The field equations, Eqs. (11.1), have to be solved subject to boundary and initial conditions.

At every point on the boundary surface $\mathcal{S}$, we have

$$x_i = \bar{x}_i \quad \text{or} \quad \sigma_i^{(n)} = \bar{\sigma}_i^{(n)}, \quad i = 1, 2, 3, \tag{11.5}$$

where the overbar ($\bar{\ }$) represents a given function.

At time $t = 0$, the location and velocity of the material particles are given as

$$x_i = X_\alpha \delta_{\alpha i}, \quad \dot{x}_i = \bar{v}_i. \tag{11.6}$$

A number of solutions have been obtained for problems in finite-strain elasticity in the static case by use of the **inverse method**. This method assumes a displacement distribution and attempts to find the corresponding stress distribution, which is in equilibrium. These results may be used in an empirical approach—in which we assume a form for the Helmholtz free energy density function and attempt to deduce the empirical constants from experiments.

11.1.1 Homogeneous Deformation

When all the strains in a body are constants, we refer to the deformation as homogeneous. Here, we assume a deformation of the form

$$x_i = A_{i\alpha} X_\alpha, \tag{11.7}$$

where $A_{i\alpha}$ is a constant, nonsingular matrix. By selecting the X_α coordinates along the principal directions of $\boldsymbol{A}$ (in which $\boldsymbol{A}$ is diagonal), we can have

$$x_1 = \lambda_1 X_1, \quad x_2 = \lambda_2 X_2, \quad x_3 = \lambda_3 X_3, \tag{11.8}$$

where λ_i are the stretches. The Green's deformation tensor is represented by the diagonal matrix

$$\boldsymbol{G} = \begin{bmatrix} \lambda_1^2 & 0 & 0 \\ 0 & \lambda_2^2 & 0 \\ 0 & 0 & \lambda_3^2 \end{bmatrix}, \tag{11.9}$$

which has, as invariants,

$$I_1 = \lambda_1^2 + \lambda_2^2 + \lambda_3^2, \quad I_2 = \lambda_1^2\lambda_2^2 + \lambda_2^2\lambda_3^2 + \lambda_3^2\lambda_1^2, \quad I_3 = \lambda_1^2\lambda_2^2\lambda_3^2. \tag{11.10}$$

As $\boldsymbol{G}$ is diagonal, no rotation is necessary, i.e., $\boldsymbol{R} = \boldsymbol{I}$, and the stress can be expressed as

$$\boldsymbol{\sigma} = a_0\boldsymbol{I} + a_1\boldsymbol{G} + a_2\boldsymbol{G}^2, \tag{11.11}$$

where

$$\begin{aligned} a_0 &= 2\rho_0\lambda_1\lambda_2\lambda_3 u_{,3}\,, \\ a_1 &= \frac{2\rho_0}{\lambda_1\lambda_2\lambda_3}[u_{,1} + (\lambda_1^2 + \lambda_2^2 + \lambda_3^2)u_{,2}\,], \\ a_2 &= -\frac{2\rho_0}{\lambda_1\lambda_2\lambda_3}u_{,2}\,. \end{aligned} \tag{11.12}$$

Explicitly, the normal components of the stress tensor are

$$\begin{aligned} \sigma_{11} &= a_0 + a_1\lambda_1^2 + a_2\lambda_1^4, \\ \sigma_{22} &= a_0 + a_1\lambda_2^2 + a_2\lambda_2^4, \\ \sigma_{33} &= a_0 + a_1\lambda_3^2 + a_2\lambda_3^4, \end{aligned} \tag{11.13}$$

and the shear components are $\sigma_{12} = \sigma_{23} = \sigma_{31} = 0$.

The equations of equilibrium are satisfied if there are no body forces or accelerations.

For an incompressible material, $I_3 = 1$, and

$$\lambda_3 = \frac{1}{\lambda_1\lambda_2}, \tag{11.14}$$

and in the expressions for the normal stresses, we replace a_0 with $-p$, the reactive hydrostatic pressure.

11.1.2 Simple Extension

An easy experiment involves stretching a circular bar in the x_1 direction. From the symmetry of the isotropic specimen, we conclude that

$$\lambda_2 = \lambda_3, \tag{11.15}$$

and the axial component of the stress $\sigma = \sigma_{11}$ is related to the axial stretch $\lambda = \lambda_1$ as

$$\sigma = a_0 + a_1\lambda^2 + a_2\lambda^4. \tag{11.16}$$

To satisfy the boundary conditions on the cylindrical surface, we have to have $\sigma_{22} = \sigma_{33} = 0$. That is,

$$a_0 + a_1\lambda_2^2 + a_2\lambda_2^4 = 0, \tag{11.17}$$

which may be viewed as an equation for $\lambda_2 = \lambda_3$ [although Eq. (11.16) for σ is coupled with this equation].

For the incompressible case, $\lambda_2^2 = 1/\lambda$, $a_0 = 0$, and we substitute $-p$ for a_0 in the preceding equation. Instead of the unknown λ_2, now we solve for p.

11.1.3 Hydrostatic Pressure

If we set $\lambda_1 = \lambda_2 = \lambda_3 = \lambda$, the state of stress corresponding to this dilatation is a hydrostatic pressure p, given by

$$p = -(a_0 + a_1\lambda^2 + a_2\lambda^4). \tag{11.18}$$

11.1.4 Simple Shear

We assume a deformation of the form

$$x_1 = X_1 + \gamma X_2, \quad x_2 = X_2, \quad x_3 = X_3, \tag{11.19}$$

where γ, the shear angle, is a constant.

From this, the deformation gradient $\boldsymbol{F}$ and the deformation tensor $\boldsymbol{G}$ are obtained as

$$\boldsymbol{F} = \begin{bmatrix} 1 & 0 & 0 \\ \gamma & 1 & 0 \\ 0 & 0 & 1 \end{bmatrix}, \quad \boldsymbol{G} = \begin{bmatrix} 1 & \gamma & 0 \\ \gamma & 1+\gamma^2 & 0 \\ 0 & 0 & 1 \end{bmatrix}. \tag{11.20}$$

The invariants of $\boldsymbol{G}$ are

$$I_1 = 3 + \gamma^2, \quad I_2 = 3 + \gamma^2, \quad I_3 = 1. \tag{11.21}$$

The stress tensor can be written as

$$\boldsymbol{\sigma} = \boldsymbol{F}^T[a_0\boldsymbol{G}^{-1} + a_1\boldsymbol{I} + a_2\boldsymbol{G}]\boldsymbol{F}. \tag{11.22}$$

After some matrix multiplications, we find

$$\begin{aligned}
\sigma_{11} &= 2\rho_0[(1+\gamma^2)u,_1 + (2+\gamma^2)u,_2 + u,_3], \\
\sigma_{22} &= 2\rho_0[u,_1 + 2u,_2 + u,_3], \\
\sigma_{33} &= 2\rho_0[u,_1 + (2+\gamma^2)u,_2 + u,_3], \\
\sigma_{12} &= 2\rho_0\gamma[u,_1 + u,_2], \\
\sigma_{13} &= \sigma_{23} = 0.
\end{aligned} \tag{11.23}$$

In linear elasticity, a block of material can be sheared with an engineering shear strain γ by an applied shear stress $\sigma_{12} = G\gamma$, where G is the modulus of rigidity. The preceding stress expressions show that, in general, in addition to the shear stress, we need normal stresses σ_{11} and σ_{22} to maintain the simple shear. This effect is called the Poynting effect. The further need for an out-of-plane stress σ_{33} to maintain the volume of the specimen is attributed to the Kelvin effect.

For an incompressible material, the preceding expressions are replaced with

$$\begin{aligned}
\sigma_{11} &= 2\rho_0\gamma^2 u_{,1}, \\
\sigma_{22} &= -2\rho_0\gamma^2 u_{,2}, \\
\sigma_{33} &= -p + 2\rho_0[u_{,1} + (2+\gamma^2)u_{,2}] = 0, \\
\sigma_{12} &= 2\rho_0\gamma[u_{,1} + u_{,2}], \\
\sigma_{13} &= \sigma_{23} = 0,
\end{aligned} \tag{11.24}$$

where we have eliminated p from σ_{11} and σ_{22} using the third equation. As the hydrostatic pressure adjusts itself to satisfy $\sigma_{33} = 0$, we do not have the Kelvin effect in this class of materials. The Poynting effect is still present. Another observation we can make is that

$$\sigma_{11} - \sigma_{22} = \gamma\sigma_{12} \tag{11.25}$$

for compressible as well as for incompressible materials.

11.1.5 Torsion of a Circular Cylinder

The inverse solution assumes the classical assumption of the torsion problem, namely, plane cross-sections of the shaft normal to its axis remain plane. Points on the cross-section undergo a rigid body rotation, the rotation angle linearly varying along the axis of the shaft. Using polar coordinates (R, θ, Z) for the initial position and (r, ϕ, z) for the final position, we have

$$r = R, \quad \phi = \theta + \alpha, \quad z = Z, \tag{11.26}$$

where $\alpha = \gamma Z$ is the angle of rotation. The Cartesian coordinates are given by

$$\begin{aligned}
x_1 &= R\cos\phi, \quad x_2 = R\sin\phi, \quad x_3 = Z, \\
X_1 &= R\cos\theta, \quad X_2 = R\sin\theta, \quad X_3 = Z.
\end{aligned} \tag{11.27}$$

We can obtain the deformation gradient (after we compute the derivatives by using the chain rule) as

$$\boldsymbol{F} = \begin{bmatrix} 1 & 0 & 0 \\ 0 & 1 & 0 \\ -\gamma R\sin\phi & \gamma R\cos\phi & 1 \end{bmatrix}. \tag{11.28}$$

It is more convenient to look at this tensor in a rotated coordinate system, with x_1' along r and x_2' along ϕ. The undeformed configuration is still described by the X_α

coordinate system. The rotation matrix is given by

$$\boldsymbol{Q} = \begin{bmatrix} \cos\phi & \sin\phi & 0 \\ -\sin\phi & \cos\phi & 0 \\ 0 & 0 & 1 \end{bmatrix}. \tag{11.29}$$

As $x'_i = Q_{ij}x_j$ and $F'_{\alpha i} = \partial_\alpha x'_i = \partial_\alpha Q_{ij}x_j$, the deformation gradient and the deformation tensors in the new coordinate system can be obtained as

$$\boldsymbol{F}' = \boldsymbol{F}\boldsymbol{Q}^T = \begin{bmatrix} \cos\phi & -\sin\phi & 0 \\ \sin\phi & \cos\phi & 0 \\ 0 & \beta & 1 \end{bmatrix},$$

$$\boldsymbol{G}' = \boldsymbol{F}'\boldsymbol{F}'^T = \begin{bmatrix} 1 & 0 & -\beta\sin\phi \\ 0 & 1 & \beta\cos\phi \\ -\beta\sin\phi & \beta\cos\phi & 1+\beta^2 \end{bmatrix}, \tag{11.30}$$

where

$$\beta = \gamma R = \gamma r. \tag{11.31}$$

The invariants are seen as

$$I_1 = 3 + \beta^2 = I_2, \quad I_3 = 1. \tag{11.32}$$

In this new system, the stress is obtained as

$$\boldsymbol{\sigma}' = \boldsymbol{F}'^T[a_0\boldsymbol{G}'^{-1} + a_1\boldsymbol{I} + a_2\boldsymbol{G}']\boldsymbol{F}'. \tag{11.33}$$

The stress components are found as

$$\begin{aligned}
\sigma'_{11} &= \sigma_{rr} = a_0 + a_1 + a_2, \\
\sigma'_{22} &= \sigma_{\phi\phi} = a_0 + (1+\beta^2)a_1 + (1 + 3\beta^2 + \beta^4)a_2, \\
\sigma'_{33} &= \sigma_{zz} = a_0 + a_1 + (1+\beta^2)a_2, \\
\sigma'_{32} &= \sigma_{z\phi} = \beta[a_1 + (2+\beta^2)a_2], \\
\sigma'_{12} &= \sigma_{r\phi} = 0, \\
\sigma'_{31} &= \sigma_{zr} = 0.
\end{aligned} \tag{11.34}$$

The equation of equilibrium in the r direction is given by

$$\frac{\partial}{\partial r}(r\sigma_{rr}) - \sigma_{\phi\phi} = 0. \tag{11.35}$$

As $\beta = \gamma r$ is a function of r, this equation cannot be satisfied in general. However, for incompressible materials (note that the deformation we have prescribed is volume preserving), a_0 is replaced with the hydrostatic pressure $-\overline{p}$, and we could have $\overline{p}$ varying along the radius to satisfy the preceding equation. That is,

$$\frac{\partial}{\partial r}(r\overline{p}) = \frac{\partial}{\partial r}[r(a_1 + a_2)] - \sigma_{\phi\phi}. \tag{11.36}$$

The torque required to create a twist of γ is obtained as

$$T = 2\pi \int_0^a r^2 \sigma_{z\phi} dr, \tag{11.37}$$

where a is the radius of the shaft. To achieve the deformation without any stretching in the axial direction, we see that an axial stress σ_{zz} is needed. This is due to the Poynting effect.

11.2 Approximate Strain Energy Functions

Because the strain energy density (Helmholtz free energy density) u is a function of I_{G1}, I_{G2}, and I_{G3}, we may attempt to expand it as a polynomial in its variables. However, these variables are not small as $I_{G1} = I_{G2} = 3$ and $I_{G3} = 1$ in the natural state. But the invariants of the Green's strain $\boldsymbol{E}$ can be small. For potential use with small strains, we expand the Helmholtz free energy in the form

$$\rho_0 u = \alpha_E I_{E1} + \frac{1}{2}(\lambda_E + 2\mu_E) I_{E1}^2 - 2\mu_E I_{E2} + \ell_E I_{E1}^3 + m_E I_{E1} I_{E2} + n_E I_{E3} + \cdots +, \tag{11.38}$$

where we have used the notation of Eringen. Note that if $I_{E1} = O(\epsilon)$ (order of ϵ, which is a small quantity), $I_{E2} = O(\epsilon^2)$ and $I_{E3} = O(\epsilon^3)$. We have kept terms up to $O(\epsilon^3)$.

We may also use the Almansi strain $\boldsymbol{e}$ to express u as a polynomial in the invariants of $\boldsymbol{e}$ as

$$\rho_0 u = \alpha_e I_{e1} + \frac{1}{2}(\lambda_e + 2\mu_e) I_{e1}^2 - 2\mu_e I_{e2} + \ell_e I_{e1}^3 + m_e I_{e1} I_{e2} + n_e I_{e3} + \cdots + . \tag{11.39}$$

For the existence of a natural state, we must have $u_{,1} = 0$. Thus $\alpha_E = 0$ and $\alpha_e = 0$. The invariants of $\boldsymbol{E}$ and $\boldsymbol{e}$ are related as the eigenvalues of the two matrices satisfy

$$E_1 = \frac{e_1}{1 - 2e_1}, \quad e_1 = \frac{E_1}{1 + 2E_1}, \quad \text{etc.} \tag{11.40}$$

Using binomial expansion and retaining terms up to $O(\epsilon^3)$, we obtain

$$E_1 = e_1 + 2e_1^2 + 4e_1^3, \quad e_1 = E_1 - 2E_1^2 + 4E_1^3. \tag{11.41}$$

The invariants can be obtained to the same degree of approximation as

$$\begin{aligned} I_{e1} &= I_{E1} + 2(I_{E1}^2 - 2I_{E2}) + 4(I_{E1}^3 - 3I_{E1}I_{E2} + 3I_{E3}) + \cdots +, \\ I_{e2} &= I_{E2} - 2I_{E1}I_{E2} + 6I_{E3} + \cdots +, \\ I_{e3} &= I_{E3} + \cdots + . \end{aligned} \tag{11.42}$$

The elastic constants in the preceding two representations are then related in the form

$$\begin{aligned} \lambda_e &= \lambda_E, \quad \mu_e = \mu_E, \quad \ell_e = \ell_E, \\ m_e &= m_E - 4\lambda_E - 12\mu_E, \quad n_e = n_E + 12\mu_E. \end{aligned} \tag{11.43}$$

The Cauchy stress and the second Piola–Kirchhoff stress are obtained from this as

$$\begin{aligned}\boldsymbol{\sigma} &= [\lambda I_{e1} + (3\ell + m_e - \lambda)I_{e1}^2 + (9m_e + n_e)I_{e2}]\boldsymbol{I} \\ &\quad + [2\mu - (m_e + n_e + 2\lambda + 2\mu)I_{e1}]\boldsymbol{e} + (n_e - 4\mu)\boldsymbol{e}^2 + \cdots +, \\ \boldsymbol{S} &= [\lambda I_{E1} + 3(\ell + m_E)I_{E1}^2 + (m_E + n_E)I_{E2}]\boldsymbol{I} \\ &\quad + [2\mu - (m_E + n_E)I_{E1}]\boldsymbol{E} + n_E\boldsymbol{E}^2 + \cdots + . \end{aligned} \tag{11.44}$$

11.2.1 Hookean Materials

The stress–strain relations previously given are explicit up to quadratic terms in the strain variables. If we set the coefficients of the quadratic and higher-order terms to zero, we get

$$\begin{aligned}\boldsymbol{\sigma} &= \lambda I_{e1}\boldsymbol{I} + 2\mu\boldsymbol{e}, \\ \boldsymbol{S} &= \lambda I_{E1}\boldsymbol{I} + 2\mu\boldsymbol{E}. \end{aligned} \tag{11.45}$$

These linear relations are called Hooke's law for isotropic materials. Materials obeying this law are called Hookean materials. The constants λ and μ are known as Lamé constants.

11.2.2 Small-Strain Approximation

If the strains and rotations are small compared with unity, the differences between $\boldsymbol{E}$ and $\boldsymbol{e}$ and those between $\boldsymbol{S}$ and $\boldsymbol{\sigma}$ vanish. We may write

$$e_{ij} = \frac{1}{2}(\partial_i u_j + \partial_j u_i). \tag{11.46}$$

The equation of linear momentum becomes

$$(\lambda + \mu)\nabla(\nabla \cdot \boldsymbol{u}) + \mu\nabla^2\boldsymbol{u} + \rho\boldsymbol{f} = \rho\boldsymbol{a}, \tag{11.47}$$

which is called the Navier equation. We may rewrite this as

$$(\lambda + \mu)u_{j,ji} + \mu u_{i,jj} + \rho f_i = \rho\ddot{u}_i. \tag{11.48}$$

The stress–strain relation

$$\sigma_{ij} = \lambda e_{kk}\delta_{ij} + 2\mu e_{ij} \tag{11.49}$$

can be inverted to give

$$e_{ij} = \frac{1}{E}[(1 + \nu)\sigma_{ij} - \nu\sigma_{kk}\delta_{ij}], \tag{11.50}$$

where E is Young's modulus and ν is Poisson's ratio, which are related to the Lamé constants in the form

$$\lambda = \frac{E\nu}{(1+\nu)(1-2\nu)}, \quad \mu = G = \frac{E}{2(1+\nu)}, \tag{11.51}$$

where G is the modulus of rigidity.

Explicitly, we have

$$\begin{aligned}
e_{11} &= \frac{1}{E}[\sigma_{11} - \nu(\sigma_{22} + \sigma_{33})], \\
e_{22} &= \frac{1}{E}[\sigma_{22} - \nu(\sigma_{33} + \sigma_{11})], \\
e_{33} &= \frac{1}{E}[\sigma_{33} - \nu(\sigma_{11} + \sigma_{22})], \\
e_{12} &= \frac{1}{2G}\sigma_{12}, \\
e_{23} &= \frac{1}{2G}\sigma_{23}, \\
e_{31} &= \frac{1}{2G}\sigma_{31}.
\end{aligned} \tag{11.52}$$

11.2.3 Plane Stress and Plane Strain

In problems involving in-plane loading of a thin plate, we assume $\sigma_{33} = \sigma_{13} = \sigma_{23} = 0$, where the x_3 axis is assumed to be perpendicular to the plate. This is called a plane stress approximation.

When the dimension of the structure in the x_3 direction is large compared with that of the other two directions, the two end sections are constrained from moving, and the applied loads in the x_1, x_2 plane do not vary with x_3, we have the plane strain approximation: $e_{33} = e_{13} = e_{23} = 0$.

In these two cases, in static elasticity of Hookean materials under small-strain approximation, the problem reduces to a two-dimensional (2D) problem. A number of exact solutions are available for simple geometric configurations. Solution techniques using complex variables developed by Kolosoff and Mushkelishvili have been extensively used on this topic.

11.3 Integrated Elasticity

In linear elasticity, we assume that (a) the stress–strain relations are linear (Hooke's law), (b) the strain–displacement relations are linear (small strains and rotations), and (c) the entire load is applied at once. In integrated elasticity or incremental elasticity, the load is increased in infinitesimal steps and for each load increment the geometry is updated. In the finite-element literature, this approach is also known as

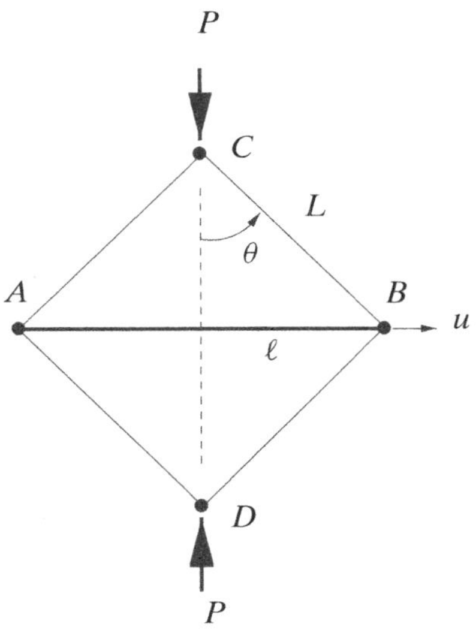

Figure 11.1. Truss with one horizontal elastic member and four rigid members.

the **updated Lagrangian method**. The incremental-loading scenario is illustrated by a simple example in the following subsection.

11.3.1 Example: Incremental Loading

Consider the five-bar truss shown in Fig. 11.1. Except for the horizontal bar AB, all the other four bars are rigid with length L, and all the joints are pin connected. The bar AB is linear elastic with Young's modulus E and constant cross-sectional area A. We start applying the load P gradually, starting when the angle $\theta = 45°$. At any angle θ, the compressive force in the member BC is given by

$$2F_1 = \frac{P}{\cos\theta},$$

and the tensile force in the horizontal, elastic member AB is

$$F = P\tan\theta.$$

The change in this force is given by

$$\Delta F = \Delta(P\tan\theta).$$

The stress increment $\Delta\sigma = \Delta F/A$, and the strain increment $\Delta\epsilon = \Delta F/EA$. The incremental strain is related to the change in the current length ℓ as

$$\Delta\epsilon = \frac{\Delta\ell}{\ell} = \frac{\Delta F}{EA}. \tag{11.53}$$

From the geometry of the truss,

$$\sin\theta = \frac{\ell}{L}, \qquad \Delta\ell = L\cos\theta\,\Delta\theta. \tag{11.54}$$

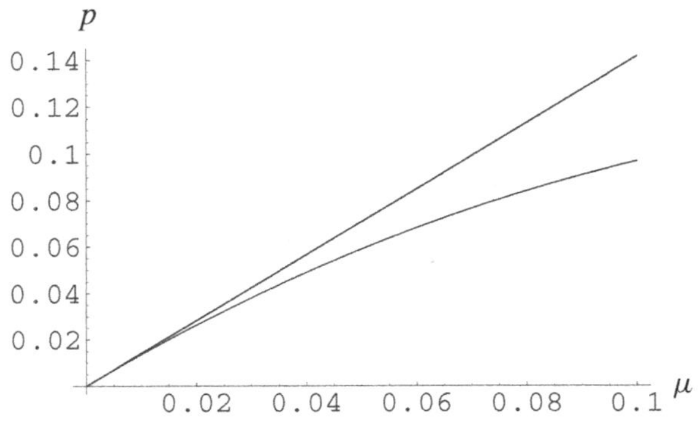

Figure 11.2. Force $p = P/EA$ as a function of $\mu = u/L$. The lower curve shows the effect of incremental change in geometry.

From Eqs. (11.53) and (11.54), we find

$$\frac{\cos\theta}{\sin\theta}\Delta\theta = \frac{\Delta F}{EA}. \tag{11.55}$$

Converting this into the differential form and integrating, we obtain

$$\frac{F}{EA} = \log\{\sqrt{2}\sin\theta\}, \tag{11.56}$$

where we have used the condition $F = 0$ when $\theta = 45^\circ$. The displacement of the bar BC is related to the angle θ in the form

$$\frac{u}{L} = \sin\theta - \frac{1}{\sqrt{2}}. \tag{11.57}$$

The applied force P and the displacement u are related as

$$p = \frac{\sqrt{2-(1+\mu)^2}}{1+\mu}\log(1+\mu), \tag{11.58}$$

where $p = P/EA$, and $\mu = \sqrt{2}u/L$.

If we neglect the change in geometry, we get

$$p = \mu. \tag{11.59}$$

A plot of the preceding two relations is shown in Fig. 11.2. It can be seen that the stiffness of the structure decreases as the angle increases.

11.4 A Variational Principle for Static Elasticity

In an isothermal process, the energy stored in a body can be expressed in terms of the Helmholtz free energy as

$$U = \int_V \rho u(I_{G1}, I_{G2}, I_{G3})dV, \tag{11.60}$$

where, based on isotropy, the energy is assumed to depend on the invariants of the Green's deformation tensor $\boldsymbol{G}$.

We consider a virtual work principle in which the virtual change in the internal energy is equal to the virtual work done by the external forces. That is,

$$\int_V \rho \delta u dV = \int_{S_\sigma} \boldsymbol{\sigma}^{(n)} \cdot \delta \boldsymbol{x} da + \int_V \boldsymbol{f} \cdot \delta \boldsymbol{x} dV, \tag{11.61}$$

where $\mathcal{S}_\sigma$ is the part of the surface on which tractions are prescribed. This can be written as

$$\int_V \rho dV \frac{2\partial u}{\partial G_{\alpha\beta}} \partial_\alpha x_i \partial_\beta \delta x_i - \int_{S_\sigma} \sigma_i^{(n)} \delta x_i da - \int_V \rho dV f_i \delta x_i = 0, \tag{11.62}$$

where we have used the symmetry of $G_{\alpha\beta}$. We have

$$\partial_\beta \delta x_i = \partial_j \delta x_i \partial_\beta x_j. \tag{11.63}$$

Using this, we rewrite the virtual work relation in the form

$$\int_V \rho dV \left[2\frac{\partial u}{\partial G_{\alpha\beta}} \partial_\alpha x_i \partial_\beta x_j \partial_j \delta x_i - f_i \delta x_i \right] - \int_{S_\sigma} \sigma_i^{(n)} \delta x_i da = 0. \tag{11.64}$$

Integrating the first term inside the volume integral by parts, we get

$$- \int_V dV \left[\partial_j \left\{ 2\frac{\rho \partial u}{\partial G_{\alpha\beta}} \partial_\alpha x_i \partial_\beta x_j \right\} + \rho f_i \right] \delta x_i$$

$$+ \int_{S_\sigma} \left\{ n_j \left[2\frac{\rho \partial u}{\partial G_{\alpha\beta}} \partial_\alpha x_i \partial_\beta x_j \right] - \sigma_i^{(n)} \right\} \delta x_i da = 0, \tag{11.65}$$

where we have set $\delta x_i = 0$ on the part of the surface where the displacements are prescribed. To satisfy this equation, inside the domain we must have

$$\partial_j \left[2\frac{\rho \partial a}{\partial G_{\alpha\beta}} \partial_\alpha x_i \partial_\beta x_j \right] + \rho f_i = 0, \tag{11.66}$$

and on the boundary,

$$n_j \left[2\frac{\rho \partial a}{\partial G_{\alpha\beta}} \partial_\alpha x_i \partial_\beta x_j \right] = \sigma_i^{(n)}. \tag{11.67}$$

The differential equation (Euler–Lagrange equation of the variational problem) can be written as

$$\partial_j \sigma_{ji} + \rho f_i = 0, \tag{11.68}$$

where σ_{ij} is defined as

$$\sigma_{ij} = 2\frac{\rho \partial u}{\partial G_{\alpha\beta}} \partial_\alpha x_i \partial_\beta x_j. \tag{11.69}$$

The boundary condition simply says

$$\sigma_i^{(n)} = n_j \sigma_{ij}. \tag{11.70}$$

The definition of the Cauchy stress and the relation between the traction and the stress tensor are exactly the ones at which we had arrived earlier.

If the applied traction and the body force are conservative, we have associated potentials Φ and φ such that

$$\sigma_i^{(n)} = -\frac{\partial \Phi}{\partial x_i}, \quad f_i = -\frac{\partial \varphi}{\partial x_i}, \tag{11.71}$$

and the virtual work principle can be written as a variational principle, namely,

$$\delta \Pi = 0, \tag{11.72}$$

where the potential energy Π is defined as

$$\Pi = \int_V \rho[u + \varphi] dV + \int_{S_\sigma} \Phi da. \tag{11.73}$$

The variational and virtual work formulations allow us to approximate the solutions of elasticity problems by assuming convenient sequences of displacement distributions. For incompressible materials, the constraint $I_{G3} = 1$ may be introduced into the variational theorem through a Lagrange multiplier function.

11.5 Isotropic Thermoelasticity

Let us assume that the coefficient of thermal expansion α is independent of strain and the total strain can be expressed as the sum of an expansion that is due to temperature change and one that is due to the state of stress. That is,

$$e_{ij} = \alpha(T - T_0)\delta_{ij} + \frac{1}{E}[(1+\nu)\sigma_{ij} - \nu\sigma_{kk}\delta_{ij}], \tag{11.74}$$

and its inverse

$$\sigma_{ij} = \lambda e_{kk}\delta_{ij} + 2\mu e_{ij} - (3\lambda + 2\mu)\alpha(T - T_0)\delta_{ij}, \tag{11.75}$$

where T_0 is a reference temperature. For a reversible, infinitesimal deformation, the differential form of the Helmholtz free energy density u can be written as

$$\begin{aligned} \rho du &= -\rho s dT + \sigma_{ij} de_{ij} \\ &= -\rho s dT + \lambda e de + 2\mu e_{ij} de_{ij} - (3\lambda + 2\mu)\alpha(T - T_0) de, \end{aligned} \tag{11.76}$$

where we have used $e = e_{kk}$.

Integrating $\rho \partial u / \partial e_{ij}$ with respect to the strains, we have

$$\rho u = \frac{\lambda e^2}{2} + \mu e_{ij} e_{ij} - (3\lambda + 2\mu)\alpha(T - T_0)e + C(T), \tag{11.77}$$

where the unknown function $C(T)$ is related to the entropy density s through

$$\begin{aligned} \rho s &= -\left(\frac{\rho \partial u}{\partial T}\right)_{e_{ij}} \\ &= -\frac{dC}{dT} - \frac{e^2}{2}\frac{d\lambda}{dT} + e\frac{d}{dT}[\alpha(3\lambda + 2\mu)(T - T_0)] - e_{ij}e_{ij}\frac{d\mu}{dT}. \end{aligned} \tag{11.78}$$

If we take $s = s(T, e_{ij})$, when $e_{ij} = 0$ the volume is constant. Then the specific heat c_v is obtained as

$$c_v = \frac{\delta Q}{\rho \delta T} = T\left(\frac{\partial s}{\partial T}\right)_{e_{ij}}$$
$$= -\frac{T}{\rho}\frac{d^2 C}{dT^2}. \quad (11.79)$$

Assuming u and s are measured from the reference state, we have

$$C(T) = -\int_{T_0}^{T} dT \int_{T_0}^{T} \rho c_v dT/T. \quad (11.80)$$

Using this, we can find the entropy density s and the internal energy:

$$\varepsilon = u + Ts. \quad (11.81)$$

If we do a similar analysis with the Gibbs free energy g, with

$$\rho dg = -\rho s dT - e_{ij} d\sigma_{ij}, \quad (11.82)$$

we can express g and s in terms of the specific heat at constant stress, c_p.

11.5.1 Specific Heats and Latent Heats

We defined two kinds of specific heats—c_v for raising the temperature while the strains are held constant and c_p while the stresses are held constant. We define the latent heat ℓ_{ij} as the amount of heat required to increase the strain e_{ij} by unity while all other strains and the temperature are fixed. Similarly, the latent heat L_{ij} is the amount of heat required to increase the stress σ_{ij} by unity while all other stresses and the temperature are fixed:

$$\delta Q = \rho T ds = c_v dT + \ell_{ij} de_{ij}$$
$$= c_p dT + L_{ij} d\sigma_{ij}. \quad (11.83)$$

From the Helmholtz free energy we have

$$\frac{\partial s}{\partial e_{ij}} = -\frac{1}{\rho}\frac{\partial \sigma_{ij}}{\partial T}. \quad (11.84)$$

Then,

$$\ell_{ij} = -\frac{T}{\rho}\frac{\partial \sigma_{ij}}{\partial T}. \quad (11.85)$$

Similarly, working with the Gibbs free energy, we have

$$\frac{\partial s}{\partial \sigma_{ij}} = -\frac{1}{\rho}\frac{\partial e_{ij}}{\partial T}, \quad (11.86)$$

$$L_{ij} = \frac{T}{\rho}\frac{\partial e_{ij}}{\partial T}. \quad (11.87)$$

For a constant-stress heat input, we may express the entropy change in two ways:

$$ds = c_v \frac{dT}{T} + \ell_{ij} \frac{de_{ij}}{dT} \frac{dT}{T} = c_p \frac{dT}{T}. \tag{11.88}$$

From this we get the relation

$$c_p - c_v = \frac{\rho}{T} L_{ij} \ell_{ij} = -\frac{T}{\rho} \left(\frac{\partial \sigma_{ij}}{\partial T} \right)_{e_{ij}} \left(\frac{\partial e_{ij}}{\partial T} \right)_{\sigma_{ij}} . \tag{11.89}$$

For the linear thermoelasticity we have

$$\ell_{ij} = -\frac{T}{\rho} \left(\frac{\partial \sigma_{ij}}{\partial T} \right)_{e_{ij}} = \frac{T}{\rho} \alpha (3\lambda + 2\mu) \delta_{ij},$$

$$L_{ij} = \frac{T}{\rho} \left(\frac{\partial e_{ij}}{\partial T} \right)_{\sigma_{ij}} = \frac{T}{\rho} \alpha \delta_{ij}, \tag{11.90}$$

$$c_p - c_v = \frac{3\alpha^2 T}{\rho} (3\lambda + 2\mu). \tag{11.91}$$

11.5.2 Strain Cooling

For linear elastic materials, the effect of stretching under adiabatic conditions is a drop in temperature. A formula was derived by Kelvin to explain this cooling effect by using

$$\delta Q = c_v dT + \ell_{ij} de_{ij} = 0 \tag{11.92}$$

to have

$$\frac{dT}{de_{ij}} = -\frac{\ell_{ij}}{c_v} = -\frac{\alpha T (3\lambda + 2\mu)}{\rho c_v} \delta_{ij}. \tag{11.93}$$

All the parameters on the right-hand side are positive, and, because of the negative sign in front, the temperature does decrease. This phenomenon is called strain cooling.

11.5.3 Adiabatic and Isothermal Elastic Modulus

From our consideration of the Helmholtz free energy, the linear thermoelastic constitutive relation is given by

$$\sigma_{ij} = \lambda e_{kk} \delta_{ij} + 2\mu e_{ij} - (3\lambda + 2\mu) \alpha (T - T_0) \delta_{ij}. \tag{11.94}$$

If the straining takes place rapidly, we may assume there is no heat flow into the specimen and we have an adiabatic process. For this,

$$c_v dT + \ell_{ij} de_{ij} = 0, \quad \ell_{ij} = \frac{T_0}{\rho} \alpha (3\lambda + 2\mu) \delta_{ij}. \tag{11.95}$$

Integrating this, we obtain

$$c_v (T - T_0) + \ell_{ij} e_{ij} = 0, \tag{11.96}$$

where the constant of integration is taken as zero by use of the reference state. From this we can solve for $T - T_0$, as

$$T - T_0 = -\ell_{ij} e_{ij} / c_v = -\frac{T_0}{\rho c_v} \alpha (3\lambda + 2\mu) e_{kk}. \tag{11.97}$$

Substituting this in the stress–strain relation, we obtain

$$\sigma_{ij} = \lambda' e_{kk} \delta_{ij} + 2\mu e_{ij}, \tag{11.98}$$

where

$$\lambda' = \lambda + \frac{T_0}{\rho c_v} \alpha^2 (3\lambda + 2\mu)^2. \tag{11.99}$$

This shows that the adiabatic value of the Lamé constant λ' is greater than its isothermal value λ.

11.5.4 Example: Rubber Elasticity

Natural rubber is known as polyisoprene. It belongs to a large group of chain molecules known as elastomers. Chain molecules are made of a repeating unit called a monomer. In addition to natural rubber, a variety of manmade elastomers are available from the chemical industry. The linear long-chain molecules can be cross-linked (typically, the linking takes place for every 100 or so monomer units) by adding sulfur. This is the vulcanization process invented by Goodyear. Now we have a tangled network in its natural state. As we apply tension, we expect the network to get less tangled, as shown in Fig. 11.3(c).

Beginning with a rubber band of cross-sectional area A_0 and length ℓ_0, we can express the total change in internal energy as

$$dE = TdS + Fd\ell, \tag{11.100}$$

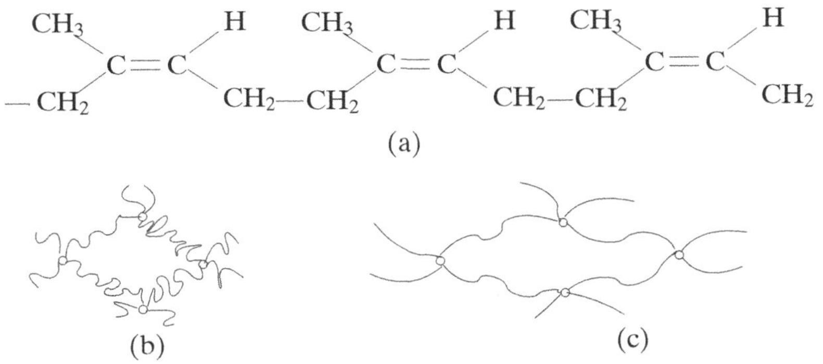

Figure 11.3. (a) Three monomer units of a polyisoprene chain, (b) a tangled cross-linked network before loading, (c) untangled network after loading (based on Noggle, 1996).

where F is the force and $d\ell$ is the change in length. It is assumed that the stretching is sufficiently slow so that no irreversible entropy production occurs. The relation between the force F and the extension $(\ell - \ell_0)$ is generally nonlinear, as indicated in Eq. (11.4). From the internal energy we get

$$\left.\frac{\partial E}{\partial S}\right|_{\ell} = T, \quad \left.\frac{\partial E}{\partial \ell}\right|_{S} = F, \tag{11.101}$$

and by considering the mixed second derivative, we get the Maxwell relation

$$\left.\frac{\partial T}{\partial \ell}\right|_{S} = \left.\frac{\partial F}{\partial S}\right|_{\ell}. \tag{11.102}$$

The Helmholtz free energy is given by

$$U = E - TS, \quad dU = -SdT + Fd\ell. \tag{11.103}$$

From this, we have

$$-S = \left.\frac{\partial U}{\partial T}\right|_{\ell}, \quad F = \left.\frac{\partial U}{\partial \ell}\right|_{T}, \tag{11.104}$$

and

$$-\left.\frac{\partial S}{\partial \ell}\right|_{T} = \left.\frac{\partial F}{\partial T}\right|_{\ell}. \tag{11.105}$$

Substituting $E - TS$ for U in the expression for F, we have

$$F = \left.\frac{\partial E}{\partial \ell}\right|_{T} - T\left.\frac{\partial S}{\partial \ell}\right|_{T}, \tag{11.106}$$

which shows the effect of decreasing entropy on the force. In other words, we are applying the force to keep the entropy down. The original, tangled network had a higher entropy or more disorder than the straightened, stretched state.

We may rewrite the preceding equation in the form

$$\left.\frac{\partial E}{\partial \ell}\right|_{T} = F + T\left.\frac{\partial S}{\partial \ell}\right|_{T} = F - T\left.\frac{\partial F}{\partial S}\right|_{\ell}. \tag{11.107}$$

Experiments involving the quantities on the right-hand side have shown that the dependence of E on ℓ is negligible. We define an ideal rubber with the property

$$\left.\frac{\partial E}{\partial \ell}\right|_{T} = 0. \tag{11.108}$$

Then, the force is given by

$$F = -T\left.\frac{\partial S}{\partial \ell}\right|_{T}, \tag{11.109}$$

which shows that the entropy decreases with length at constant temperature. Another assumption regarding ideal rubber is that the volume is preserved during elongation, that is,

$$V = V_0. \tag{11.110}$$

If we pull the rubber band sufficiently fast, we can assume no heat input into the material. Then,

$$c_v dT - F d\ell = 0, \tag{11.111}$$

$$\left.\frac{\partial T}{\partial \ell}\right|_S = \frac{F}{c_v} = -\frac{T}{c_v}\left.\frac{\partial S}{\partial \ell}\right|_T. \tag{11.112}$$

Because the partial derivative is negative, we find a temperature rise in the rubber band.

In our Mooney model, with $\lambda = \ell/\ell_0$,

$$G_{11} = \lambda^2, \quad G_{22} = G_{33} = \frac{1}{\lambda}, \tag{11.113}$$

$$I_{G1} = \lambda^2 + \frac{2}{\lambda}, \quad I_{G2} = 2\lambda + \frac{1}{\lambda^2}. \tag{11.114}$$

For the ideal rubber, we assume the Mooney–Rivlin constant, A_2 in Eq. (11.4), is zero, and, then,

$$U = \rho_0 V_0 A_1 \left(\lambda^2 + \frac{2}{\lambda} - 3\right). \tag{11.115}$$

The force becomes

$$F = \frac{\partial U}{\partial \ell} = \frac{2\rho_0 V_0}{\ell_0} A_1 \left(\lambda - \frac{1}{\lambda^2}\right). \tag{11.116}$$

The second Piola–Kirchhoff stress S_{11} is given by

$$S_{11} = 2\rho_0 A_1 \left(\lambda - \frac{1}{\lambda^2}\right). \tag{11.117}$$

It has been shown by use of statistical mechanics that the constant A_1 has the value

$$A_1 = \frac{RT}{2zM}, \tag{11.118}$$

where R is the universal gas constant, T is the absolute temperature, z is the number of monomer units between cross-links, and M is the molecular weight of the monomer unit.

11.6 Linear Anisotropic Materials

When a material is anisotropic, the linear relation between the stress and strain can be written as

$$\sigma_{ij} = S_{ijkl} e_{kl}, \quad e_{ij} = C_{ijkl} \sigma_{kl}, \tag{11.119}$$

where S is called the stiffness tensor and C is the compliance tensor. As the stress and strain tensors are symmetric, we note that

$$S_{ijkl} = S_{jikl} = S_{ijlk}, \quad C_{ijkl} = C_{jikl} = C_{ijlk}. \tag{11.120}$$

The Helmholtz free energy density for this material has the form

$$\rho u = \frac{1}{2} S_{ijkl} e_{ij} e_{kl}, \tag{11.121}$$

which shows a further symmetry:

$$S_{ijkl} = S_{klij}. \tag{11.122}$$

The same symmetry exists for the compliance tensor C.

It is customary (and convenient) to use the Voigt notation, which associates a linear array with a matrix, such as

$$\begin{Bmatrix} \sigma_{11} \\ \sigma_{22} \\ \sigma_{33} \\ \sigma_{23} \\ \sigma_{13} \\ \sigma_{12} \end{Bmatrix} = \begin{Bmatrix} \sigma_1 \\ \sigma_2 \\ \sigma_3 \\ \sigma_4 \\ \sigma_5 \\ \sigma_6 \end{Bmatrix}, \tag{11.123}$$

to write

$$\sigma_i = S_{ij} e_j, \quad e_i = C_{ij} \sigma_j. \tag{11.124}$$

Now, the stiffness and compliance matrices are symmetric, 6×6 matrices, with 21 independent constants in each. Using material symmetry, such as orthotropy, transverse isotropy, and isotropy, we find that these constants reduce to 9, 5, and 2, respectively.

11.7 Invariant Integrals

In elasticity problems involving singularities, the invariant integral for small strains and rotations (small deformation gradients) presented in Chapter 10 has found numerous applications. These singularities usually include crack tips with small plastically deformed regions or inhomogeneities extending with a stationary field in the moving coordinate system. As mentioned earlier, when the closed surface surrounds a singularity, the integral may be nonzero, and we may write

$$\Gamma_i = \int_S [\rho \varepsilon n_i - \sigma_{kj} \partial_i u_j n_k] dS. \tag{11.125}$$

For isothermal cases, we use the Helmholtz free energy density u for the internal energy ε to obtain

$$\Gamma_i = \int_S [\rho u n_i - \sigma_{kj} \partial_i u_j n_k] dS. \tag{11.126}$$

Recalling that the singularity is moving through the medium with velocity components V_i, we see that material particles are crossing the surface S with velocity $(-\boldsymbol{V})$. Then, $\Gamma_i V_i$ is the energy absorbed per unit time by the singularity, and Γ_i have dimensions of a force. This force is known as a material force or Eshelby force. In the plane strain case, for a crack tip moving in the x_1 direction, we have

$$J = \int_S [\rho u n_1 - \sigma_{kj}\partial_1 u_j n_k] dS, \tag{11.127}$$

which is the well-known J integral introduced by Rice.

Instead of singular points, we may also consider lines of singularities or surfaces of singularities in the context of invariant integrals.

SUGGESTED READING

Antman, S. (2004). *Nonlinear Problems in Elasticity*, 2nd ed., Springer.
Barber, J. R. (2002). *Elasticity*, 2nd ed., Kluwer Academic.
Green, A. E. and Zerna, W. (1968). *Theoretical Elasticity*, 2nd ed., Clarendon.
Landau, L. D. and Lifshitz, E. M. (1986). *Theory of Elasticity*, 3rd ed., Pergamon.
Noggle, J. H. (1996). *Physical Chemistry*, 3rd ed., Harper Collins.
Ogden, R. W. (1984). *Nonlinear Elastic Deformation*, Ellis Horwood.
Timoshenko, S. P. and Goodier, J. N. (1970). *Theory of Elasticity*, 3rd ed., McGraw-Hill.

EXERCISES

11.1. A circular cylindrical shaft of radius a and length ℓ is made of a Mooney material. One end of the shaft is rotated by an angle β with respect to the other end. Compute the torque T and the axial force N required for twisting the shaft under the assumptions of rigid body rotation of each cross-section and no change in axial length.

11.2. Obtain the balance of linear momentum for static elasticity for plane stress and plane strain:

$$\partial_1\sigma_{11} + \partial_2\sigma_{12} + \rho f_1 = 0,$$

$$\partial_1\sigma_{12} + \partial_2\sigma_{22} + \rho f_2 = 0.$$

Assume that the body forces are conservative and that they are obtained from a potential φ in the form

$$\rho f_1 = -\partial_1\varphi, \quad \rho f_2 = -\partial_2\varphi.$$

Show that the Airy stress function ϕ, introduced in the form

$$\sigma_{11} = \partial_2^2\phi + \varphi, \quad \sigma_{22} = \partial_1^2\phi + \varphi, \quad \sigma_{12} = -\partial_1\partial_2\phi,$$

satisfies the balance of linear momentum.
Using these in the stress–strain relations for a Hookean material, obtain the governing equations for ϕ for plane stress and plane strain. When $f_i = 0$, show that, for both plane stress and plane strain, ϕ satisfies the biharmonic equation

$$\nabla^4\phi = 0.$$

11.3. Transform the biharmonic equation in the preceding problem into the polar coordinates r and θ. Obtain the functions $F_n(r)$ in the separable solutions

$$\phi = F_n(r)e^{in\theta},$$

where i is the imaginary number and n is an integer.

11.4. By considering the dimensional changes of a parallelepiped, show that for infinitesimal strains the volume change per unit volume is given by

$$\frac{\Delta V}{V} = e = e_{ii}.$$

Using Hooke's law, relate this to the hydrostatic pressure

$$p = -\frac{1}{3}\sigma_{ii}$$

in the form

$$e = -\frac{p}{K},$$

where K is the bulk modulus. Express K in terms of E and ν and, also, in terms of the Lamé constants λ and μ.

11.5. Consider linear, plane elasticity in the x_1, x_2 plane. Show that the J integral defined by

$$J = \oint_S [\rho u n_1 - \sigma_{kj} \partial_1 u_j n_k] dS,$$

where S is now a curve, is invariant with respect to any two choices, $S = C_1$ and $S = C_2$, as long as the stresses are nonsingular in the region between the two contours.

11.6. In the Navier equations, neglecting the body forces, assume that

$$u = \sin k(x \pm c_L t), \quad v = 0, \quad w = 0,$$

where k is the wave number and c is the wave speed. Obtain an expression for c_L in terms of the Lamé constants and the density. Waves with displacements oriented in the direction of propagation are called longitudinal or P waves.

11.7. In the preceding exercise, if we assume that

$$u = 0, \quad v = \sin k(x \pm c_T t), \quad w = 0,$$

we have shear (transverse) or S waves. Obtain the wave speed in this case.

11.8. In the Navier equations (without body forces), assume that

$$\boldsymbol{u} = \nabla\phi + \nabla \times \boldsymbol{\psi},$$

where ϕ and $\boldsymbol{\psi}$ are known as the scalar and vector potentials. Show that both of these functions satisfy the wave equation, with the wave speed c_L for ϕ and c_T for $\boldsymbol{\psi}$. Also show that the ϕ wave represents volume change and the $\boldsymbol{\psi}$ wave represents rotations.

11.9. In linear thermoelasticity, assuming that λ, μ, α, and the specific heats c_v and c_p are independent of temperature T, obtain the densities of Helmholtz free energy u, internal energy ε, entropy s, enthalpy h, and the Gibbs free energy, g.

11.10. Compute the ratio E'/E, where E' and E are, respectively, the adiabatic and the isothermal Young's moduli.

11.11. Assume a rubber band is under constant tension. Show that its length decreases according to

$$\left.\frac{\partial \ell}{\partial T}\right|_F = \frac{1}{k}\left.\frac{\partial S}{\partial \ell}\right|_T,$$

where

$$k = \left.\frac{\partial F}{\partial \ell}\right|_T$$

when it is uniformly heated.

12 Fluid Dynamics

Fluid dynamics is another active area of research because of its impact on aerodynamics, flows in turbomachinery, design of propellers, and applications in convective heat transfer. Our aim here is to discuss the basic concepts of fluid dynamics in the general context of continuum mechanics. In Chapter 10 we briefly discussed Stokes fluids. Here we continue with some solutions by using the inverse method. The Newtonian fluid is considered next. We conclude with ideal fluids and thermodynamic implications.

12.1 Basic Equations

The equations of conservation of mass and the balance of momenta and the constitutive relations are

$$\begin{aligned}
\frac{\partial \rho}{\partial t} + \partial_i(\rho v_i) &= 0, \\
\partial_i \sigma_{ij} + \rho f_j &= \rho \dot{v}_j, \\
\sigma_{ij} &= \sigma_{ji}, \\
\sigma_{ij} &= (-p + a_0)\delta_{ij} + a_1 d_{ij} + a_2 d_{ik} d_{kj}, \quad \text{compressible}, \\
\sigma_{ij} &= -\bar{p}\delta_{ij} + a_1 d_{ij} + a_2 d_{ik} d_{kj}, \quad \text{incompressible},
\end{aligned} \tag{12.1}$$

where the phenomenological coefficients a_i are functions of the invariants of the deformation rate tensor, i.e.,

$$a_i = a_i(I_1, I_2, I_3), \quad a_0(0,0,0) = 0, \tag{12.2}$$

p is the thermodynamic pressure, and $\bar{p}$ is the reactive hydrostatic pressure. All the invariants pertain to the deformation rate tensor $\boldsymbol{d}$. For the incompressible case we have $I_1 = 0$.

For nonnegative entropy production we must have

$$a_0 I_1 + a_1(I_1^2 - 2I_2) + a_2(I_1^3 - 3I_1 I_2 + 3I_3) > 0, \tag{12.3}$$

which reduces to

$$-2a_1 I_2 + 3a_2 I_3 > 0 \tag{12.4}$$

for the incompressible case.

We also have the defining kinematic relations

$$\dot{v}_i = \frac{\partial v_i}{\partial t} + v_j \partial_j v_i. \tag{12.5}$$

The thermodynamic pressure p is known from an equation of state,

$$p = p(\rho, T). \tag{12.6}$$

These equations are supplemented with a system of boundary and initial conditions. At every point on the surface, three quantities have to be prescribed. These could be three components of the traction vector, three components of the velocity vector, or a mixture involving traction components and their complementary velocity components. At solid boundaries the velocity of the fluid should be equal to the velocity of the solid. This is known as the **no-slip condition**.

12.2 Approximate Constitutive Relations

The constitutive relations of Stokes fluids require specification of three functions a_i, known as the phenomenological coefficients, which are functions of the three invariants of the deformation rate tensor **d**, as mentioned earlier. We may assume $I_1 = O(\epsilon)$, $I_2 = O(\epsilon^2)$, $I_3 = O(\epsilon^3)$, and expand the phenomenological coefficients. To include quadratic terms in ϵ in the stress expressions, let

$$\begin{aligned} a_0 &= \lambda I_1 + \lambda_1 I_1^2 + \lambda_2 I_2, \\ a_1 &= 2\mu + 2\mu_1 I_1, \\ a_2 &= 4\nu, \end{aligned} \tag{12.7}$$

where we have used the same constants as in Eringen. Of course, the possibility that there are real fluids with the preceding phenomenological constants is not precluded even if I_1 is not a small quantity. These fluids are called second-order fluids.

The stresses are obtained for compressible fluids as

$$\boldsymbol{\sigma} = [(-p + \lambda)I_1 + \lambda_1 I_1^2 + \lambda_2 I_2]\boldsymbol{I} + 2(\mu + \mu_1 I_1)\boldsymbol{d} + 4\nu \boldsymbol{d}^2, \tag{12.8}$$

where the coefficient of $\boldsymbol{I}$ has to be replaced with $-\bar{p}$ for the incompressible case. Interestingly, the entropy production inequality shows that, for the incompressible case, $\nu = 0$. Thus there is no second-order incompressible fluid.

12.3 Newtonian Fluids

We consider a class of fluids for which the stresses are linear functions of the deformation rate. These are called Newtonian fluids. From the second-order constitutive relations we remove all second-order terms to have

$$a_0 = \lambda I_1, \quad a_1 = 2\mu, \quad a_2 = 0, \tag{12.9}$$

and

$$\sigma_{ij} = (\lambda I_1 - p)\delta_{ij} + 2\mu d_{ij}, \quad \text{compressible},$$
$$\sigma_{ij} = -\bar{p}\delta_{ij} + 2\mu d_{ij}, \quad \text{incompressible}. \tag{12.10}$$

The constants λ and μ are called the dilatational and shear viscosities. The entropy production rate is proportional to

$$\sigma_{ij} d_{ij} = \lambda I_1^2 + 2\mu d_{ij} d_{ij} = \lambda I_1^2 + 2\mu(I_1^2 - 2I_2). \tag{12.11}$$

To assess the constraints on the constants, let us express the invariants in terms of the eigenvalues of $\boldsymbol{d}$, namely, the stretchings d_i:

$$I_1 = d_1 + d_2 + d_3, \quad I_2 = d_1 d_2 + d_2 d_3 + d_3 d_1. \tag{12.12}$$

Now,

$$\sigma_{ij} d_{ij} = (\lambda + \frac{2}{3}\mu)(d_1 + d_2 + d_3)^2 + 2\mu[(d_1 - d_2)^2 + (d_2 - d_3)^2 + (d_3 - d_1)^2]. \tag{12.13}$$

For nonnegative dissipation we must have

$$\lambda + \frac{2}{3}\mu \geq 0, \quad \mu \geq 0. \tag{12.14}$$

The hydrostatic pressure for compressible Newtonian fluids can be found as

$$-\frac{1}{3}\sigma_{ii} = p - (\lambda + \frac{2}{3}\mu) I_1, \tag{12.15}$$

and for incompressible fluids,

$$-\frac{1}{3}\sigma_{ii} = \bar{p}. \tag{12.16}$$

If

$$\lambda + \frac{2}{3}\mu = 0, \tag{12.17}$$

we see that the hydrostatic pressure and the thermodynamic pressure are equal for the compressible Newtonian fluids. This condition is known as the Stokes condition. This approximation is often used for water and air.

By defining a deviatoric deformation rate,

$$d'_{ij} = d_{ij} - \frac{1}{3} d_{kk} \delta_{ij}, \tag{12.18}$$

we can have

$$\sigma_{ij} = -p\delta_{ij} + 2\mu d'_{ij} \tag{12.19}$$

for Newtonian fluids under the Stokes assumption.

12.4 Inviscid Fluids

From the first-order theory, previously described, by eliminating the linear terms in d_{ij}, we obtain a zeroth-order theory, where

$$\sigma_{ij} = -p\delta_{ij}. \tag{12.20}$$

These are called inviscid fluids. The thermodynamic pressure p is dependent on the density ρ and temperature T:

$$p = p(\rho, T). \tag{12.21}$$

If p is independent of temperature, the fluid is called a barotropic fluid.

For incompressible zeroth-order fluids,

$$\sigma_{ij} = -\bar{p}\delta_{ij}, \tag{12.22}$$

and such a fluid is known as an ideal fluid. The pressure here is a reactive quantity.

As mentioned earlier, let us consider a few classic flows by using the inverse method. In the inverse method, we prescribe a velocity distribution and see if the balance of momentum can be satisfied.

12.5 Shearing Flow

We assume the velocities are of the form

$$v_1 = u(y), \quad v_2 = v_3 = 0, \tag{12.23}$$

where we use $y = x_2$ and $x = x_1$. For the time being, u is arbitrary. As shown in Fig 12.1, the flow is confined between a solid static wall at $y = 0$ and a free surface or a rigid plate at $y = h$. The deformation rate tensor is given by

$$\boldsymbol{d} = \begin{bmatrix} 0 & u'/2 & 0 \\ u'/2 & 0 & 0 \\ 0 & 0 & 0 \end{bmatrix}, \tag{12.24}$$

which has the invariants

$$I_1 = 0, \quad I_2 = -(u'/2)^2, \quad I_3 = 0, \tag{12.25}$$

where $u' = du/dy$. The stress is obtained as

$$\sigma = (a_0 - p)\begin{bmatrix} 1 & 0 & 0 \\ 0 & 1 & 0 \\ 0 & 0 & 1 \end{bmatrix} + \frac{a_1 u'}{2}\begin{bmatrix} 0 & 1 & 0 \\ 1 & 0 & 0 \\ 0 & 0 & 0 \end{bmatrix} + \frac{a_2 (u')^2}{4}\begin{bmatrix} 1 & 0 & 0 \\ 0 & 1 & 0 \\ 0 & 0 & 0 \end{bmatrix}, \tag{12.26}$$

where

$$a_i = a_i[0, -(u'/2)^2, 0] = a_i(y). \tag{12.27}$$

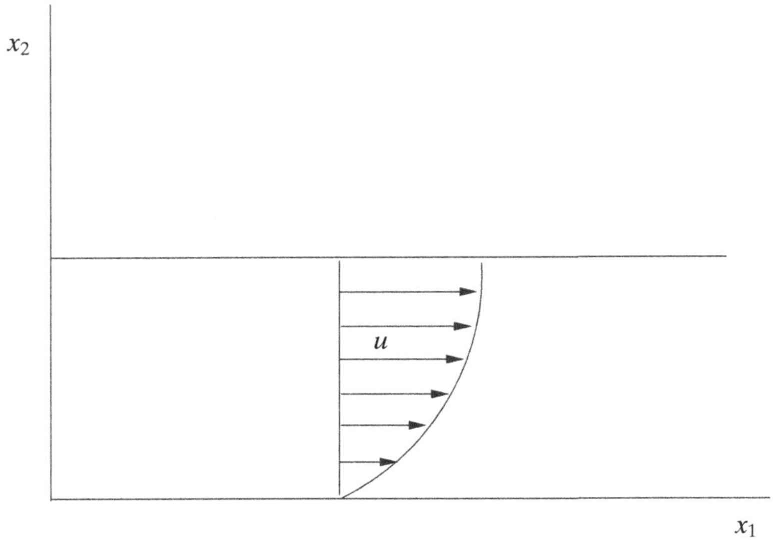

Figure 12.1. Shearing flow over a fixed surface.

The nonzero stresses are

$$\sigma_{11} = \sigma_{22} = (a_0 - p) + \frac{1}{4}a_2(u')^2,$$
$$\sigma_{33} = a_0 - p,$$
$$\sigma_{12} = \frac{1}{2}a_1 u'. \tag{12.28}$$

Let us neglect the body force. For a steady flow, the acceleration

$$\dot{v}_i = \frac{\partial v_i}{\partial t} + v_j \partial_j v_i = 0. \tag{12.29}$$

The equations of linear momentum give

$$-\partial_1 p + [\frac{1}{2}a_1 u']' = 0,$$
$$[-p + a_0 + \frac{1}{4}a_2(u')^2]' = 0, \tag{12.30}$$

where a prime indicates differentiation with respect to $x_2 = y$. Integrating the second equation gives

$$-p + a_0 + \frac{1}{4}a_2(u')^2 = g(x), \tag{12.31}$$

where $g(x)$ has to be found. Substituting this into the first of Eqs. (12.30), we obtain

$$\frac{dg}{dx} + (\frac{1}{2}a_1 u')' = 0. \tag{12.32}$$

This equation is satisfied (by separation of variables) if

$$\frac{dg}{dx} = -C, \quad \left(\frac{1}{2}a_1 u'\right)' = C, \tag{12.33}$$

where C is a constant. Integrating these, we obtain

$$g(x) = -Cx + D, \quad a_1 u' = 2Cy + E, \tag{12.34}$$

where D and E are new constants. The stresses can be expressed as

$$\sigma_{11} = \sigma_{22} = -Cx + D, \quad \sigma_{12} = Cy + E/2. \tag{12.35}$$

We may consider two situations: (a) There is a free surface at $y = h$, and (b) there is a rigid plate at $y = h$ moving with a velocity U.

Case (a): The constants $C = 0$, $D = 0$ for $\sigma_{22} = 0$ for all x and $E = 0$ for $\sigma_{12} = 0$ at $y = h$. The velocity distribution is obtained from the nonlinear differential equation

$$a_1 u' = 0, \quad u(0) = 0, \tag{12.36}$$

which allows $u(y) = 0$ as a trivial solution. Recall that the coefficient a_1 is a function of the unknown u'.

Case (b): The differential equation

$$a_1 u' = 2Cy + E \tag{12.37}$$

may have solutions.

For a Newtonian fluid, $a_1 = 2\mu$, and

$$u' = \frac{1}{2\mu}(2Cy + E), \quad u = \frac{1}{2\mu}(Cy^2 + Ey + G), \tag{12.38}$$

where the constant $G = 0$ to have $u(0) = 0$. If we stipulate σ_{11} is independent of x, we get $C = 0$. This gives the simple linear distribution for the velocity:

$$u(y) = Uy/h\,. \tag{12.39}$$

In addition, if there is a pressure drop in the x direction,

$$\partial_1 \sigma_{11} = -C, \quad u(y) = \frac{Uy}{h} + \frac{C}{2\mu}y(y - h). \tag{12.40}$$

12.6 Pipe Flow

The axisymmetric flow in the direction of the axis of a uniform circular pipe is called a Hagen–Poiseuille flow. This case is better suited for cylindrical coordinates, r, θ, and z. Because of the axisymmetry, all quantities are independent of θ. We assume a velocity distribution of the form

$$v_r = 0, \quad v_\theta = 0, \quad v_z = w(r). \tag{12.41}$$

The boundary condition on the pipe wall is

$$w(a) = 0, \tag{12.42}$$

where a is the radius of the pipe.

Using the gradient operator

$$\nabla = \mathbf{e}_r \frac{\partial}{\partial r} + \mathbf{e}_\theta \frac{\partial}{r\partial \theta} + \mathbf{e}_z \frac{\partial}{\partial z}, \tag{12.43}$$

and

$$\boldsymbol{d} = \frac{1}{2}[\nabla \boldsymbol{v} + (\nabla \boldsymbol{v})^T], \tag{12.44}$$

we find

$$\boldsymbol{d} = \begin{bmatrix} d_{rr} & d_{r\theta} & d_{rz} \\ d_{r\theta} & d_{\theta\theta} & d_{\theta z} \\ d_{rz} & d_{\theta z} & d_{zz} \end{bmatrix} = \frac{1}{2} w' \begin{bmatrix} 0 & 0 & 1 \\ 0 & 0 & 0 \\ 1 & 0 & 0 \end{bmatrix}. \tag{12.45}$$

The invariants are

$$I_1 = 0, \quad I_2 = -\frac{1}{4}(w')^2, \quad I_3 = 0, \tag{12.46}$$

and the functional dependence of the phenomenological coefficients can be seen as

$$a_i = a_i[0, -(w')^2/4, 0] = a_i(r). \tag{12.47}$$

The nonzero stresses are given by

$$\sigma_{rr} = \sigma_{zz} = -p + a_0 + \frac{1}{4} a_2 (w')^2, \quad \sigma_{\theta\theta} = -p + a_0, \quad \sigma_{rz} = \frac{1}{2} a_1 w'. \tag{12.48}$$

The equations of motion are

$$\frac{\partial(r\sigma_{rr})}{\partial r} - \sigma_{\theta\theta} = 0,$$

$$\frac{\partial(r\sigma_{rz})}{\partial r} - r\frac{\partial \sigma_{zz}}{\partial z} = 0. \tag{12.49}$$

Substituting the stress expressions, we find that the equations of motion give

$$-\frac{\partial p}{\partial r} + a_0' + \frac{1}{4r}[a_2 r (w')^2]' = 0,$$

$$-\frac{\partial p}{\partial z} + \frac{1}{2r}[a_1 r w']' = 0. \tag{12.50}$$

Integrating the first equation, we have

$$-p + a_0 + \int_0^r \frac{1}{4r}[a_2 r (w')^2]' \, dr = f(z), \tag{12.51}$$

where f is an unknown function. Substituting this in the second equation, we obtain

$$\frac{df}{dz} + \frac{1}{2r}[a_1 r w']' = 0. \tag{12.52}$$

This equation is satisfied if

$$\frac{df}{dz} = C, \quad \frac{1}{2r}[a_1 r w']' = -C, \tag{12.53}$$

where the constant C has to be found. From these,

$$f = Cz + D, \quad a_1 r w' = -Cr^2 + E, \quad a_1 w' = -Cr + \frac{E}{r}. \tag{12.54}$$

For w' to be finite at $r = 0$, we must have $E = 0$. Then the differential equation for w is given by

$$a_1 w' = -Cr, \tag{12.55}$$

which has to be integrated with the condition $w(a) = 0$. The stresses are obtained as

$$\begin{aligned}
\sigma_{rr} &= \sigma_{zz} = Cz + D - \int_0^r \frac{1}{4r} a_2 (w')^2 dr, \\
\sigma_{\theta\theta} &= Cz + D - \frac{a_2 (w')^2}{4} - \int_0^r \frac{1}{4r} a_2 (w')^2 dr, \\
\sigma_{rz} &= -\frac{1}{2} Cr.
\end{aligned} \tag{12.56}$$

For the Newtonian fluid,

$$w' = -\frac{C}{2\mu} r, \quad w = \frac{C}{4\mu}(a^2 - r^2), \tag{12.57}$$

which is the well-known parabolic distribution of the velocity. The stresses are

$$\begin{aligned}
\sigma_{rr} &= \sigma_{zz} = \sigma_{\theta\theta} = Cz + D, \\
\sigma_{rz} &= -\frac{1}{2} Cr.
\end{aligned} \tag{12.58}$$

For a second-order fluid, $a_1 = 2\mu + 2\mu_1 I_1 = 2\mu$ and $a_2 = 4\nu$. Then

$$w(r) = \frac{C}{4\mu}(a^2 - r^2), \tag{12.59}$$

and

$$\begin{aligned}
\sigma_{rr} &= \sigma_{zz} = Cz + D - \frac{C^2 \nu r^2}{8\mu^2}, \\
\sigma_{\theta\theta} &= Cz + D - \frac{3C^2 \nu r^2}{8\mu^2}, \\
\sigma_{rz} &= -\frac{1}{2} Cr.
\end{aligned} \tag{12.60}$$

The hydrostatic pressure can be found to be

$$\bar{p} = -Cz - D + \frac{5C^2 \nu r^2}{24\mu^2}. \tag{12.61}$$

The pressure difference between a point on the wall and another on the axis is seen as

$$\bar{p}|_a - \bar{p}|_0 = \frac{5C^2 \nu a^2}{24\mu^2}, \tag{12.62}$$

which isolates the effect of the second-order constant ν. At the exit of a pipe, this pressure difference would cause the flow to "swell" as it encounters atmospheric pressure on the outer surface. This an indicator of a higher-order fluid. The constant C can be found from the pressure drop or from the volume flow rate,

$$Q = \int_0^a 2\pi r v_z dr = \frac{\pi C a^4}{8\mu}. \tag{12.63}$$

12.7 Rotating Flow

The flow in a circular cylindrical container with the velocity field

$$v_r = 0, \quad v_\theta = r\omega(r), \quad v_z = 0 \tag{12.64}$$

is known as the Couette flow. Again, in polar coordinates, the deformation rate can be written as

$$\boldsymbol{d} = \begin{bmatrix} 0 & \beta & 0 \\ \beta & 0 & 0 \\ 0 & 0 & 0 \end{bmatrix}, \tag{12.65}$$

where we use

$$\beta = \frac{r\omega'}{2}. \tag{12.66}$$

The invariants are found as

$$I_1 = 0, \quad I_2 = -\beta^2, \quad I_3 = 0. \tag{12.67}$$

The phenomenological coefficients are

$$a_i = a_i(0, -\beta^2, 0) = a_i(r). \tag{12.68}$$

The stress tensor in polar coordinates is obtained as

$$\boldsymbol{\sigma} = (a_0 - p)\begin{bmatrix} 1 & 0 & 0 \\ 0 & 1 & 0 \\ 0 & 0 & 1 \end{bmatrix} + a_1\beta \begin{bmatrix} 0 & 1 & 0 \\ 1 & 0 & 0 \\ 0 & 0 & 0 \end{bmatrix} + a_2\beta^2 \begin{bmatrix} 1 & 0 & 0 \\ 0 & 1 & 0 \\ 0 & 0 & 0 \end{bmatrix}. \tag{12.69}$$

Explicitly, the nonzero stresses are

$$\sigma_{rr} = \sigma_{\theta\theta} = a_0 - p + a_2\beta^2,$$

$$\sigma_{zz} = a_0 - p,$$

$$\sigma_{r\theta} = a_1\beta. \tag{12.70}$$

Because of the centrifugal acceleration, we have the equations of motion,

$$\partial_r \sigma_{rr} = -\rho r \omega^2, \tag{12.71a}$$

$$\partial_r (r^2 \sigma_{r\theta}) = 0, \tag{12.71b}$$

$$\partial_z \sigma_{zz} = -\rho g, \tag{12.71c}$$

where we have assumed axisymmetry and a body force (weight of the fluid) in the z direction which is assumed to be downward.

Integrating Eq. (12.71b), we get

$$\sigma_{r\theta} = \frac{C}{r^2}, \quad a_1 \beta = \frac{C}{r^2}, \quad a_1 \omega' = \frac{2C}{r^3}, \tag{12.72}$$

where C is a constant. The second relation verifies that C is indeed a constant and not a function of z. The third relation is the nonlinear differential equation for $\omega(r)$.

The remaining two equations of motion give

$$-\partial_r p + (a_0 + a_2 \beta^2)' = -\rho r \omega^2, \tag{12.73a}$$

$$-\partial_z p = -\rho g. \tag{12.73b}$$

Integrating Eq (12.73b), we obtain,

$$p = \rho g z + F(r), \tag{12.74}$$

where we must find $F(r)$ by substituting in Eq. (12.73a). This gives

$$F' = \rho r \omega^2 + (a_0 + a_2 \beta^2)'. \tag{12.75}$$

Integrating this relation gives

$$F(r) = \int \rho r \omega^2 dr + a_0 + a_2 \beta^2 + D, \tag{12.76}$$

$$p = \rho g z + \int \rho r \omega^2 dr + a_0 + a_2 \beta^2 + D. \tag{12.77}$$

The stresses are found as

$$\sigma_{rr} = \sigma_{\theta\theta} = -\rho g z - \int \rho r \omega^2 dr - D,$$

$$\sigma_{zz} = -\rho g z - \int \rho r \omega^2 dr - a_2 \beta^2 - D,$$

$$\sigma_{r\theta} = \frac{C}{r^2}. \tag{12.78}$$

The singularity in the shear stress at $r = 0$ can be avoided if $C = 0$ or if the flow is confined between two concentric cylinders. Let us assume the inner cylinder has a radius of a and the outer cylinder has a radius of b. Further, we may take the outer cylinder as fixed and the inner cylinder as rotating with an angular velocity of Ω.

The torque required for the rotation is given by

$$T = \int_0^{2\pi} \sigma_{r\theta} r^2 d\theta|_{r=a} = 2\pi C. \tag{12.79}$$

If there is a free surface, the normal stress on it must be zero. That is, $\sigma_{zz} = 0$. This gives

$$z = -\frac{1}{\rho g}\left[\rho \int r\omega^2 dr + a_2\beta^2 + D\right]. \tag{12.80}$$

Let us illustrate this with a second-order fluid. Because the invariant $I_1 = 0$, we have

$$a_1 = 2\mu, \quad a_2 = 4\nu. \tag{12.81}$$

Then,

$$\omega' = \frac{C}{\mu r^3}, \quad \omega = \frac{C}{2\mu}\left[\frac{1}{b^2} - \frac{1}{r^2}\right], \tag{12.82}$$

where we used $\omega(b) = 0$. The condition $\omega(a) = \Omega$ gives

$$C = 2\mu\Omega / \left[\frac{1}{b^2} - \frac{1}{a^2}\right] = \frac{T}{2\pi}. \tag{12.83}$$

The integral

$$\int_b^r \rho r\omega^2 dr = \frac{\rho C^2}{2b^2}\left[\frac{r^2}{b^2} - \frac{b^2}{r^2} - 4\ln\frac{r}{b}\right]. \tag{12.84}$$

The stresses are obtained as

$$\begin{aligned}
\sigma_{rr} &= \sigma_{\theta\theta} = -\rho g z - D - \frac{\rho C^2}{2b^2}\left[\frac{r^2}{b^2} - \frac{b^2}{r^2} - 4\ln\frac{r}{b}\right],\\
\sigma_{zz} &= -\rho g z - D - \frac{\rho C^2}{2b^2}\left[\frac{r^2}{b^2} - \frac{b^2}{r^2} - 4\ln\frac{r}{b}\right] - \frac{C^2\nu}{\mu^2 r^4},\\
\sigma_{r\theta} &= \frac{C}{r^2}.
\end{aligned} \tag{12.85}$$

To maintain the velocity distribution assumed for the Couette flow, the normal stress, σ_{zz}, has to be applied. As we have seen, if there is a free surface, the equation for the surface is given by

$$z = -\frac{1}{g}\left\{\frac{D}{\rho} + \frac{C^2}{2b^2}\left[\frac{r^2}{b^2} - \frac{b^2}{r^2} - 4\ln\frac{r}{b}\right] + \frac{C^2\nu}{\rho\mu^2 r^4}\right\}, \tag{12.86}$$

where the constant D can be selected to have $z = 0$ at any chosen radius. Remembering that the z axis points downward, the last term involving ν contributes to a rise in the fluid level. This rise is higher at the inner wall than at the outer wall. This effect is known as the Weissenberg effect. Usually this effect is demonstrated by inserting a rotating rod into a second-order fluid in a cylindrical vessel to see the fluid rise around the rod.

12.8 Navier–Stokes Equations

For Newtonian fluids we have

$$\sigma_{ij} = (-p + \lambda I_1)\delta_{ij} + 2\mu d_{ij}, \quad \text{compressible}, \tag{12.87}$$

$$\sigma_{ij} = -\bar{p}\delta_{ij} + 2\mu d_{ij}, \quad \text{incompressible}, \tag{12.88}$$

where

$$I_1 = \partial_k v_k, \quad d_{ij} = \frac{1}{2}(\partial_i v_j + \partial_j v_i). \tag{12.89}$$

Substituting these in the equations of motion, we have

$$\rho(\frac{\partial v_i}{\partial t} + v_j \partial_j v_i) = \rho f_i - \partial_i p + \partial_i (\lambda \partial_j v_j) + \partial_j[\mu(\partial_i v_j + \partial_j v_i)], \tag{12.90}$$

$$\rho(\frac{\partial v_i}{\partial t} + v_j \partial_j v_i) = \rho f_i - \partial_i \bar{p} + \partial_j[\mu(\partial_i v_j + \partial_j v_i)], \tag{12.91}$$

for compressible and incompressible fluids, respectively. These are the well-known Navier–Stokes equations.

Next, let us consider the incompressible and compressible cases separately.

12.9 Incompressible Flow

The density is constant for incompressible flows, and the viscosity may depend on temperature. Assuming isothermal conditions, the Navier–Stokes equations become

$$\frac{\partial v_i}{\partial t} + v_j \partial_j v_i = f_i - \frac{1}{\rho}\partial_i \bar{p} + \nu \partial_j \partial_j v_i, \tag{12.92}$$

where we used $\partial_j v_j = 0$, and ν is the kinematic viscosity defined as

$$\nu = \frac{\mu}{\rho}. \tag{12.93}$$

Further, for steady flows, we get

$$v_j \partial_j v_i = f_i - \frac{1}{\rho}\partial_i \bar{p} + \nu \partial_j \partial_j v_i. \tag{12.94}$$

We may scale this equation by using a velocity U and a length scale L, representing the distance along which significant changes in quantities take place. Let us use the following scaled replacements:

$$v_i \to v_i/U, \quad \bar{p} \to \bar{p}/\rho U^2, \quad f_i \to f_i/U^2, \quad x_i \to x_i/L \tag{12.95}$$

The equations of motion can be written as

$$\frac{\partial v_i}{\partial t} + v_j \partial_j v_i = f_i - \partial_i \bar{p} + \frac{1}{R}\partial_j \partial_j v_i, \tag{12.96}$$

where the nondimensional constant R is given by

$$R = \rho U L/\mu. \tag{12.97}$$

This number is called the Reynolds number, which is a measure of the ratio of the inertia forces to viscous forces. The Reynolds number plays an important role in the transition of laminar flows into turbulent flows.

12.10 Compressible Flow

Equation of motion (12.90) has to be supplemented with the equation of conservation of mass,

$$\frac{\partial \rho}{\partial t} + \partial_i(\rho v_i) = 0. \tag{12.98}$$

The thermodynamic pressure is given by the equation of state,

$$p = p(\rho, T). \tag{12.99}$$

With pressure and temperature variations, we need the energy equation

$$\rho \dot{\varepsilon} = \sigma_{ij} d_{ij} + \partial_i q_i + \rho h \tag{12.100}$$

to supplement the balance and the conservation relations.

Assuming an ideal gas with Fourier's law of heat conduction, we find

$$\dot{\varepsilon} = c_v \dot{T}, \quad p = R\rho T, \quad q_i = k\partial_i T. \tag{12.101}$$

Using these, we obtain

$$\rho c_v \dot{T} - \partial_i(k\partial_i T) - \rho h = (-R\rho T + \lambda d_{kk})d_{ii} + 2\mu d_{ij} d_{ij}. \tag{12.102}$$

The three equations of motions for v_i have to be solved with the conservation of mass for ρ and the energy equation for T, simultaneously.

12.11 Inviscid Flow

When the dilatational viscosity λ and the shear viscosity μ are negligible we call the fluid an inviscid fluid. The equations previously discussed for compressible fluids become

$$\frac{\partial v_i}{\partial t} + v_j \partial_j v_i = f_i - \frac{1}{\rho}\partial_i p,$$

$$\frac{\partial \rho}{\partial t} + \partial_i(\rho v_i) = 0, \tag{12.103}$$

$$\rho c_v \dot{T} + R\rho T \partial_i v_i - \partial_i(k\partial_i T) - \rho h = 0,$$

$$p = p(\rho, T). \tag{12.104}$$

The system of corresponding equations for incompressible fluids is

$$\frac{\partial v_i}{\partial t} + v_j \partial_j v_i = f_i - \frac{1}{\rho} \partial_i \bar{p},$$

$$\partial_i v_i = 0,$$

$$\rho c \dot{T} - \partial_i (k \partial_i T) - \rho h = 0, \tag{12.105}$$

where c is the specific heat. In the incompressible case, the specific heats at constant volume and at constant pressure are the same. The heat conduction equation can be solved separately in this case.

12.11.1 Speed of Sound

Equations (12.103) for inviscid, compressible fluids can be applied to small disturbances in a still fluid of density ρ_0 kept at a constant pressure p_0 and temperature T_0. The small deviations from the quiescent state can be written as

$$v_i = 0 + v_i', \quad p = p_0 + p', \quad \rho = \rho_0 + \rho'. \tag{12.106}$$

Neglecting the body forces and the quadratic and higher-order terms, Eqs. (12.103) give

$$\rho_0 \frac{\partial v_i'}{\partial t} = -\partial_i p' = -\left.\frac{\partial p}{\partial \rho}\right|_{\rho_0} \frac{\partial \rho'}{\partial x_i}, \tag{12.107}$$

$$\frac{\partial \rho'}{\partial t} = -\rho_0 \frac{\partial v_i'}{\partial x_i}. \tag{12.108}$$

Eliminating the velocity components, we obtain the wave equation for ρ',

$$\frac{\partial^2 \rho'}{\partial t^2} = \left.\frac{\partial p}{\partial \rho}\right|_{\rho_0} \frac{\partial^2 \rho'}{\partial x_i \partial x_i}. \tag{12.109}$$

A wave of density fluctuation moving in the direction $\boldsymbol{n}$ can be represented as

$$\rho' = A e^{ik(n_i x_i - ct)}, \tag{12.110}$$

where k and c represent the wave number and the wave speed. From the wave equation, we find

$$c^2 = \left.\frac{\partial p}{\partial \rho}\right|_{\rho_0}. \tag{12.111}$$

Noting

$$p' = c^2 \rho', \tag{12.112}$$

we see that the velocity components v_i' and the pressure p' also satisfy the preceding wave equation.

12.11.2 Method of Characteristics

In Eqs. (12.103) for inviscid compressible fluids, if the thermal gradients are absent, the flow is known as an isentropic flow. We briefly examine the one-dimensional isentropic flow. The thermodynamic pressure can be assumed to be a function of the density,

$$p = p(\rho), \tag{12.113}$$

and the only component of velocity is u in the x direction. Thus,

$$\rho\frac{\partial u}{\partial t} + \rho u\frac{\partial u}{\partial x} + c^2\frac{\partial \rho}{\partial x} = 0, \tag{12.114}$$

$$\frac{\partial \rho}{\partial t} + u\frac{\partial \rho}{\partial x} + \rho\frac{\partial u}{\partial x} = 0, \tag{12.115}$$

where

$$c^2 = \left.\frac{\partial p}{\partial \rho}\right|_{\rho}. \tag{12.116}$$

These are two nonlinear partial differential equations that come under the classification of quasi-linear equations because of the presence of the highest derivatives in the linear form. One way to solve such systems is through the use of the method of characteristics, provided we have a hyperbolic system of equations. Because the linear wave equation with constant coefficients allows solutions in terms of characteristic variables $x - \bar{c}t$ and $x + \bar{c}t$, we begin by assuming that there are characteristic curves $s(x, t)$, such that

$$ds = dx - \bar{c}dt, \quad \frac{\partial}{\partial x} = \frac{\partial}{\partial s}, \quad \frac{\partial}{\partial t} = -\bar{c}\frac{\partial}{\partial s}. \tag{12.117}$$

Along these curves we can replace $\partial u/\partial t$ with $-\bar{c}\partial u/\partial x$. The isentropic flow equations give

$$\rho(-\bar{c} + u)u_x + c^2\rho_x = 0, \tag{12.118}$$

$$\rho u_x + (-\bar{c} + u)\rho_x = 0, \tag{12.119}$$

where $u_x = \partial u/\partial x$, etc. Setting the determinant of this homogeneous system to zero, we find

$$\bar{c} = u + c, \quad \bar{c} = u - c, \tag{12.120}$$

which show real slopes for the two sets of characteristic curves—a requirement for hyperbolic systems. Further, the eigenvectors of the system show that

$$\rho u_x + c\rho_x = 0, \quad \text{along} \quad \frac{dx}{dt} = u - c, \tag{12.121}$$

$$\rho u_x - c\rho_x = 0, \quad \text{along} \quad \frac{dx}{dt} = u + c. \tag{12.122}$$

These relations can be integrated to obtain the so-called Riemann invariants,

$$u + \int_{\rho_0}^{\rho} \frac{c d\rho}{\rho} = C_1, \tag{12.123}$$

$$u - \int_{\rho_0}^{\rho} \frac{c d\rho}{\rho} = C_2, \tag{12.124}$$

which are constants along the characteristic curves.

12.12 Bernoulli Equation

When the flow is steady, we may obtain a conservation principle known as the Bernoulli equation under certain conditions. The convective acceleration term can be written as

$$v_j \partial_j v_i = v_j(\partial_j v_i - \partial_i v_j) + v_j \partial_i v_j = 2v_j w_{ji} + \frac{1}{2}\partial_i(v^2). \tag{12.125}$$

We assume there is a potential φ for the (conservative) body force $\boldsymbol{f}$ such that

$$f_i = -\frac{\partial \varphi}{\partial x_i}. \tag{12.126}$$

We also assume the fluid is either incompressible or barotropic, that is, $p = p(\rho)$. Then we have a function P such that

$$\frac{1}{\rho}\partial_i(p, \bar{p}) = \frac{\partial P}{\partial x_i}. \tag{12.127}$$

With these, the equations of motion for an inviscid fluid can be expressed as

$$\nabla(\varphi + P + \frac{1}{2}v^2) = -2\boldsymbol{v} \cdot \boldsymbol{w}. \tag{12.128}$$

Now, if the flow is irrotational, $\boldsymbol{w}$ is zero, and

$$\varphi + P + \frac{1}{2}v^2 = \text{constant}. \tag{12.129}$$

Even for a rotational flow, if we dot multiply Eq. (12.128) by $\boldsymbol{v}$ and use

$$\boldsymbol{v} \cdot (\boldsymbol{v} \cdot \boldsymbol{w}) = 0, \tag{12.130}$$

we get

$$v_i \partial_i(\varphi + P + \frac{1}{2}v^2) = 0. \tag{12.131}$$

Because the stream lines are parallel to $\boldsymbol{v}$, this equation shows that the directional derivative of the total potential along a stream line is zero. That is,

$$\varphi + P + \frac{1}{2}v^2 = \text{constant along a stream line}. \tag{12.132}$$

12.13 Invariant Integrals

Following our discussion of the invariant surface integrals in Chapter 10, the momentum flux exerts a force

$$F_j = \int_S [\rho v_i v_j - \sigma_{ij}] n_i dS \tag{12.133}$$

on a singularity in a fluid flow. As an example, consider a rotating cylinder (or a vortex line) perpendicular to the plane of a two-dimensional (2D) flow with far-field velocity U in the x direction. Assuming the center of the cylinder is located at the origin $x = 0$, $y = 0$, the velocity distribution is

$$v_x = U - \frac{\Gamma y}{2\pi r^2}, \quad v_y = \frac{\Gamma x}{2\pi r^2}, \tag{12.134}$$

where Γ is the circulation due to the rotating cylinder. For an incompressible fluid,

$$\sigma_{ij} = -\bar{p}\delta_{ij} = -[p_0 - \frac{1}{2}v^2]\delta_{ij}, \tag{12.135}$$

where p_0 is the stagnation pressure. When this is inserted into the integral taken from 0 to $2\pi r$, the p_0 term vanishes and

$$F_x = \frac{1}{2}\rho \int_0^{2\pi r} [2(v_x n_x + v_y n_y)v_x - v^2 n_x] ds, \tag{12.136}$$

$$F_y = \frac{1}{2}\rho \int_0^{2\pi r} [2(v_x n_x + v_y n_y)v_y - v^2 n_y] ds. \tag{12.137}$$

For the circle, $n_x = x/r$ and $n_y = y/r$, and

$$v_x n_x + v_y n_y = U n_x. \tag{12.138}$$

After integration, we get the famous result known as the Kutta–Joukowski theorem or the Magnus effect in the case of a rotating cylinder:

$$F_1 = 0, \quad F_2 = \rho \Gamma U. \tag{12.139}$$

The invariant integral involving the internal energy has applications in compressible fluids.

SUGGESTED READING

Batchelor, G. K. (1967). *An Introduction to Fluid Dynamics*, Cambridge University Press.
Fredrickson, A. G. (1964). *Principles and Applications of Rheology*, Prentice-Hall.
Gidaspow, D. (1994). *Multiphase Flow and Fluidization*, Academic.
Levich, V. G. (1962). *Physicochemical Hydrodynamics*, Prentice-Hall.
Panton, R. L. (2005). *Incompressible Flow*, Cambridge University Press.
Schlischting, H. (1955). *Boundary Layer Theory*, McGraw-Hill.

EXERCISES

12.1. A fluid has the phenomenological coefficients

$$a_0 = 0, \quad a_1 = -2\mu I_2, \quad a_2 = 0.$$

It is contained between two parallel plates separated by a distance h. Calculate the velocity distribution and the flow rate if the rate of pressure drop is a constant.

12.2. Assuming the preceding fluid is flowing through a pipe of radius a, obtain the velocity profile and the flow rate for a given rate of pressure drop.

12.3. For the rotating flow between two cylinders, assume the inner radius $a = b/4$, where b is the outer radius. Further, choose the arbitrary constant D such that the stress-free surface has zero height at $r/b = 1$. Introducing a nondimensional parameter for z, obtain a plot of z versus r/b. Use an appropriate nondimensional group to represent the coefficient ν. Use a coordinate system with the z axis pointing downward.

12.4. An incompressible fluid obeys the constitutive law

$$d_{12} = \frac{1}{2\mu}\begin{cases}\sigma_{12} & \text{if} \quad |\sigma_{12}| > \tau \\ 0 & \text{if} \quad |\sigma_{12}| \le \tau\end{cases},$$

where τ is a constant. Assuming it is flowing between two fixed parallel plates at $x_2 = \pm a$ and the only nonzero velocity is $u = v_1$ caused by a pressure gradient in the x_1 direction, obtain the velocity profile, the domain in which $\boldsymbol{d} = \boldsymbol{0}$, and the flow rate in terms of the pressure gradient. Such flows are known as "plug flows" in the literature. This type of constitutive relation describes a special case of Bingham fluids, which are distinguished by a critical shear stress below which the rate of deformation is zero.

12.5. In a 2D flow in a plane normal to the x_3 axis, if a Newtonian fluid is incompressible and irrotational, we have

$$\partial_1 v_1 + \partial_2 v_2 = 0, \quad \partial_2 v_1 - \partial_1 v_2 = 0.$$

If the velocity potential ϕ and the stream function ψ are defined as

$$v_1 = \partial_1\phi = \partial_2\psi, \quad v_2 = \partial_2\phi = -\partial_1\psi,$$

obtain the governing equations for ϕ and ψ.
Show that the contour lines on which ϕ is a constant are perpendicular to the contour lines on which ψ is a constant. Also show that the velocity vector is tangential to the ψ contours.

12.6. The equation of state for a van der Waals gas is

$$p = \frac{RT}{v - b} - \frac{a}{v^2}, \quad v = \frac{1}{\rho}.$$

Show that the energy equation can be written for this case as

$$\rho c_v \dot{T} + \frac{\rho RT}{1 - b\rho} d_{ii} - \partial_i(k\partial_i T) - \rho h = \lambda(d_{ii})^2 + 2\mu d_{ij} d_{ij}.$$

12.7. An incompressible, Newtonian fluid occupies the domain $-\infty < x < \infty, 0 < y < \infty$. At time $t = 0$, the boundary $y = 0$ is given a velocity U. Assuming that it is a plane flow and there are no body forces or pressure drop, obtain the velocity distribution as a function of time and location.

12.8. In the preceding exercise, instead of a suddenly applied velocity for the boundary $y = 0$, assume the boundary is oscillating with the time dependence, $u(x, 0) = U \cos \omega t$. Neglecting the transient effects, obtain the velocity distribution as a function of t and y.

12.9. As a second modification to Exercise 12.7, assume there is a second, nonmoving boundary at $y = h$ and the boundary $y = 0$ is oscillating as before. Again, obtain velocity distribution.

12.10. A rod of radius a occupies the region $-\infty < x < \infty, 0 < r < a$. It is surrounded by an incompressible, Newtonian fluid. At $t = 0$, the rod is given a velocity U in the axial direction. To find the velocity of the fluid follow these instructions:

(a) Show that in cylindrical coordinates the axial velocity u satisfies

$$\frac{\partial u}{\partial t} = \nu \left[\frac{\partial^2 u}{\partial r^2} + \frac{1}{r} \frac{\partial u}{\partial r} \right].$$

(b) After taking the Laplace transform

$$\bar{u}(r, p) = \int_0^\infty u(r, t) e^{-pt} \, dt,$$

obtain the solution $\bar{u}$ in terms of the modified Bessel function K_0 as

$$\bar{u}(r, p) = A K_0(\sqrt{p/\nu}\, r),$$

and the other solution, I_0, is not useful as it is unbounded as $r \to \infty$.

(c) Choose the constant A to satisfy the boundary condition at $r = a$.

(d) In the Laplace inversion integral,

$$u(r, t) = \frac{1}{2\pi i} \int_\Gamma \bar{u}(r, p) e^{pt} \, dp,$$

where Γ is a vertical line in the complex p plane, just right of the origin, deform the contour to go from $-\infty$ to 0 just below the branch cut along the negative real axis and from 0 to $-\infty$ just above the branch cut. Note that the branch cut is necessary as K_0 is not single valued without it.

(e) Using the substitutions $p = \nu\lambda^2 e^{i\pi}$ for the integration above the branch cut and $p = \nu\lambda^2 e^{-i\pi}$ for the integration below the branch cut, obtain

$$u(r, t) = \frac{2U}{\pi} \int_0^\infty \frac{J_0(\lambda r) Y_0(\lambda a) - Y_0(\lambda r) J_0(\lambda a)}{J_0^2(\lambda a) + Y_0^2(\lambda a)} \frac{e^{-\nu\lambda^2 t}}{\lambda} \, d\lambda.$$

(f) Obtain the viscous force exerted by the fluid on a unit length of the rod.

12.11. In the case of one-dimensional isentropic flow, if

$$p = A\rho^\gamma,$$

where A is a constant, obtain the characteristic slopes and the Riemann invariants.

12.12. Equations (12.114) and (12.115) can be linearized by the so-called hodograph transformation in which we use u and ρ as independent variables and x and t as dependent variables. Obtain the linearized equations in this way.

12.13. In a plane flow with far-field velocity $v_x = U$, there is a line source of strength q at the origin (the line being normal to the plane of the flow). The velocity distribution that is due to the source is

$$v_x = \frac{qx}{2\pi r^2}, \qquad v_y = \frac{qy}{2\pi r^2}.$$

Find the force on the source using the invariant momentum flux.

13 Viscoelasticity

Materials in which the stress at a point depends on the entire history of the strain are called viscoelastic materials. Hyperelastic materials and Stokes fluids can be considered as special cases of viscoelastic materials. In general, viscoelastic materials exhibit a combination of solidlike and fluidlike characteristics. Examples of viscoelastic materials can be found in plastics, polymers, and metals at high temperatures. The nonlinear viscoelastic behavior of metals is commonly referred to as **creep**. This chapter mainly deals with linear viscoelastic materials. Before we consider general constitutive relations, it is of interest from a historical perspective to examine some one-dimensional classic models. As in the case of fluids, the behavior of viscoelastic materials under hydrostatic pressure differs from their behavior under shear stresses. This experimental evidence has to be incorporated when we generalize the one-dimensional behavior to three dimensions. We also refer to viscoelastic solids and viscoelastic fluids on the basis of whether the displacement under a constantly applied load reaches a finite limit.

13.1 Kelvin–Voigt Solid

The one-dimensional force–displacement relation for the Kelvin–Voigt solid is written as

$$P = ku + \mu \dot{u}, \tag{13.1}$$

where k and μ are material constants, P is the uniaxial force, and u and $\dot{u}$ are the infinitesimal displacement and displacement rate, respectively.

A mechanical model representing this constitutive relation has a spring with stiffness k and a **dashpot** with viscosity μ. This arrangement is shown in Fig. 13.1. Because of the linear nature of this force–displacement relation, we can obtain the response of the material to a given history of the load by superposing the response of the material to a load in the form of a **unit step function**. This superposition principle is applicable to all linear viscoelastic materials. We will demonstrate this using the Laplace transform later in this chapter. The response u of a viscoelastic material to a unit load input is called the **creep compliance function** $J(t)$. For the Kelvin–Voigt

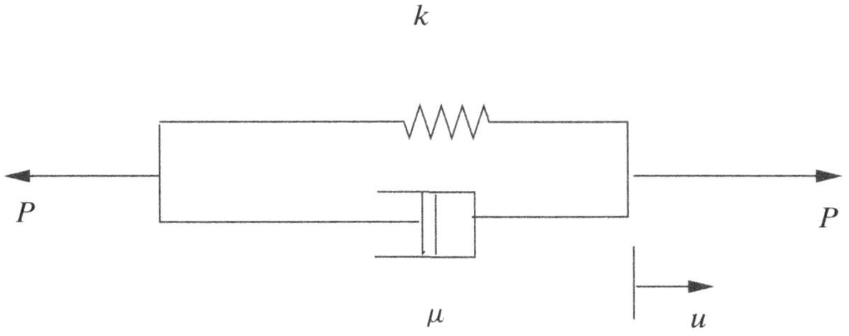

Figure 13.1. Mechanical model for a Kelvin–Voigt solid.

solid, we have to solve the differential equation

$$\dot{u} + \frac{k}{\mu}u = \frac{1}{\mu}H(t), \tag{13.2}$$

where H is the Heaviside step function. Using the initial condition $u(0) = 0$, we find that the solution of the differential equation is

$$J(t) = \frac{1}{k}[1 - e^{-t/\tau}], \quad \tau = \frac{\mu}{k}, \tag{13.3}$$

where the constant τ is called the relaxation time. For high values of the viscosity, the relaxation time can be seen to be large.

The load history needed to produce a constant displacement of unity is called the **stress relaxation function** $G(t)$. Substituting

$$u(t) = H(t), \tag{13.4}$$

into differential equation (13.1), we get

$$P(t) = kH(t) + \mu\delta(t), \tag{13.5}$$

where $\delta(t)$ is the Dirac delta function. Figure 13.2 shows sketches of the creep compliance and stress relaxation functions for a Kelvin–Voigt material.

13.2 Maxwell Fluid

The mechanical model for a Maxwell fluid consists of a spring and dashpot arranged in a series, as shown in Fig. 13.3. The load displacement relation is given by

$$\dot{u} = \frac{\dot{P}}{k} + \frac{P}{\mu}, \tag{13.6}$$

where, again, k and μ are constants. Because of the series arrangement, for a constant applied load the material can flow to an unlimited extent, given enough time. The creep compliance and stress relaxation functions for the Maxwell fluid are

$$J(t) = [\frac{1}{k} + \frac{t}{\mu}]H(t), \quad G(t) = ke^{-t/\tau}, \quad \tau = \frac{\mu}{k}. \tag{13.7}$$

Figure 13.4 shows sketches of these functions.

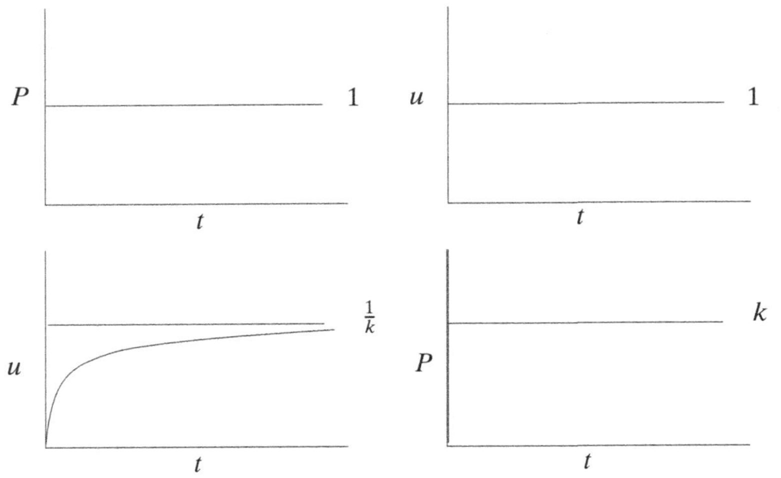

Figure 13.2. Creep compliance and stress relaxation functions for a Kelvin–Voigt solid.

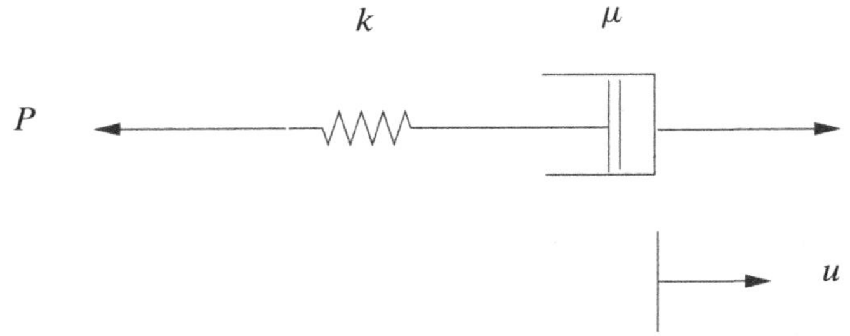

Figure 13.3. Mechanical model for a Maxwell fluid.

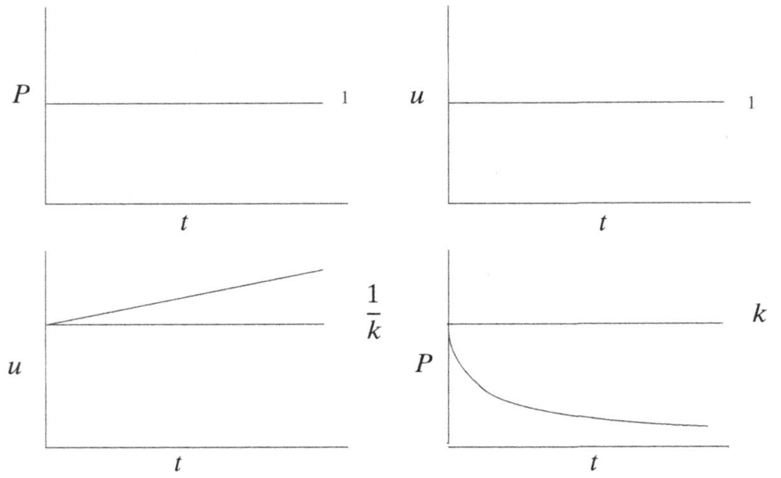

Figure 13.4. Creep compliance and stress relaxation functions for a Maxwell fluid.

13.3 Standard Linear Solid

Based on the mechanical model shown in Fig. 13.5, we have the constitutive relation

$$\dot{P} + pP = k[\dot{u} + qu], \tag{13.8}$$

where

$$p = \frac{k_2}{\mu}, \quad k = k_1 + k_2, \quad q = \frac{k_1 k_2}{\mu(k_1 + k_2)}. \tag{13.9}$$

An alternative, three-parameter mechanical model that has the same constitutive relation as in Eq. (13.8) is given in the Exercises.

With the initial conditions $u(0^-) = 0$ or $P(0^-) = 0$, we can solve the differential equation to obtain the creep compliance or the stress relaxation function. Also, from the symmetry of the differential equation from the creep compliance function, by changing the parameters we can obtain the stress relaxation function. For example, to obtain the creep compliance, we set $P(t) = H(t)$ and the differential equation becomes

$$\dot{u} + qu = \frac{1}{k}[\delta(t) + pH(t)]. \tag{13.10}$$

Integrating this equation from $(0 - \epsilon)$ to $(0 + \epsilon)$ and letting $\epsilon \to 0$, we obtain

$$u(0^+) = \frac{1}{k}. \tag{13.11}$$

For $t > 0$ we have

$$\dot{u} + qu = \frac{p}{k}, \tag{13.12}$$

which has the solution

$$u(t) = \frac{p}{qk} + Ce^{-qt}. \tag{13.13}$$

Using the initial condition, Eq. (13.11), we find that

$$C = \frac{1}{k}[1 - \frac{p}{q}], \tag{13.14}$$

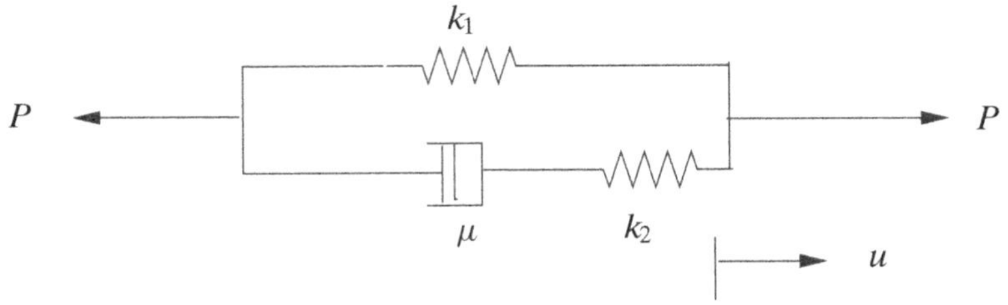

Figure 13.5. Mechanical model for a standard linear solid.

and $J(t) = u(t)$ is given by

$$J(t) = \frac{1}{k}[(1 - \frac{p}{q})e^{-qt} + \frac{p}{q}]. \quad (13.15)$$

The stress relaxation function is obtained by the replacements

$$p \to q, \quad q \to p, \quad k \to 1/k \quad (13.16)$$

in $J(t)$ as

$$G(t) = k[(1 - \frac{q}{p})e^{-pt} + \frac{q}{p}]. \quad (13.17)$$

The creep compliance and stress relaxation functions show that the stress relaxation time $(1/p)$ and the strain relaxation time $(1/q)$ are generally distinct. We define the relaxation times

$$\tau_s = \frac{1}{p}, \quad \tau_e = \frac{1}{q} \quad (13.18)$$

for the stress and strain, respectively. The creep compliance and stress relaxation functions can be written as

$$J(t) = \frac{1}{k}\left[(1 - \frac{\tau_e}{\tau_s})e^{-t/\tau_e} + \frac{\tau_e}{\tau_s}\right], \quad (13.19)$$

$$G(t) = k\left[(1 - \frac{\tau_s}{\tau_e})e^{-t/\tau_s} + \frac{\tau_s}{\tau_e}\right]. \quad (13.20)$$

Sketches of these functions when $\tau_e > \tau_s$ are shown in Fig. 13.6.

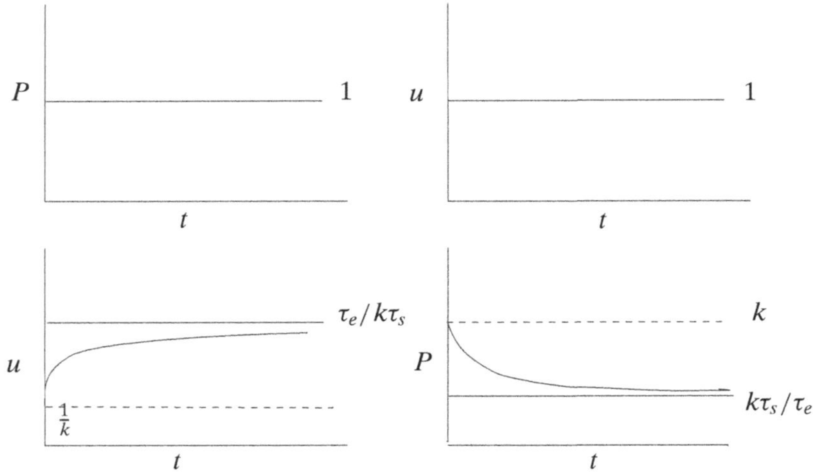

Figure 13.6. Creep compliance and stress relaxation functions for a standard linear solid when $\tau_e > \tau_s$.

13.4 Superposition Principle

Consider the case in which an arbitrary load history is given as shown in Fig. 13.7. Assume we have the creep compliance function for the material. As shown in the displacement diagram, if the load were kept constant at $t = \tau$ the displacement at t would be $u(t)$. Because of a step increase in the load at τ, the increment in u is given by

$$du(t) = J(t - \tau)dP(\tau). \tag{13.21}$$

Integrating this, we obtain

$$u(t) = \int_0^t J(t - \tau)\frac{dP}{d\tau}d\tau. \tag{13.22}$$

Similarly, if the displacement history is known, the load history can be found with the stress relaxation function $G(t)$, as

$$P(t) = \int_0^t G(t - \tau)\frac{du}{d\tau}d\tau. \tag{13.23}$$

These integral representations are known as the Boltzmann superposition integrals. Taking the Laplace transform of these convolution integrals, Eqs. (13.22) and (13.23), with transform variable s, we get

$$\mathcal{L}[u] = \mathcal{L}[J]s\mathcal{L}[P], \quad \mathcal{L}[P] = \mathcal{L}[G]s\mathcal{L}[u] \tag{13.24}$$

or

$$s^2\mathcal{L}[J]\mathcal{L}[G] = 1. \tag{13.25}$$

This relation shows the reciprocal property of the two functions, rate of creep compliance, and rate of stress relaxation in the Laplace domain.

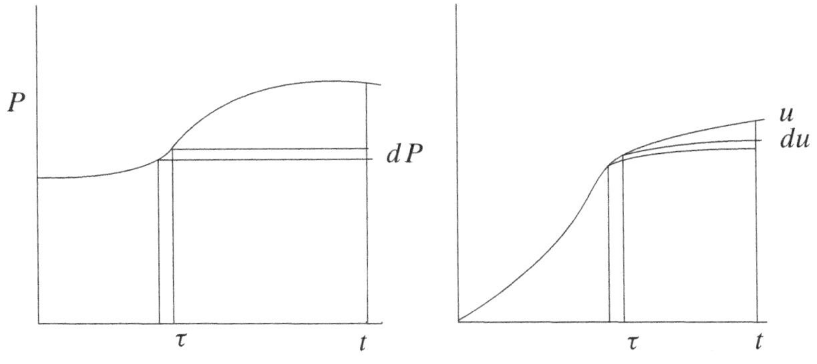

Figure 13.7. Superposition of strain by use of the creep compliance function.

13.5 Constitutive Laws in the Operator Form

With the notation

$$D = \frac{d}{dt}, \tag{13.26}$$

and

$$\mathcal{P}(D) = p_m D^m + p_{m-1} D^{m-1} + \cdots + p_0, \quad \mathcal{Q}(D) = q_n D^n + q_{n-1} D^{n-1} + \cdots + q_0, \tag{13.27}$$

the load displacement relation can be written for a multiparameter model as

$$\mathcal{P}(D) P(t) = \mathcal{Q}(D) u(t). \tag{13.28}$$

This representation is known as the operator form of the constitutive relation. Taking the Laplace transform, with $t \to s$ and the assumption that the initial values of P and all its time derivatives up to the order of $(m-1)$ and of u and its derivatives up to the order of $(n-1)$ are zero,

$$\mathcal{P}(s)\overline{P} = \mathcal{Q}(s)\bar{u}, \tag{13.29}$$

where we have used the notation

$$\overline{P} = \mathcal{L}[P, t \to s]. \tag{13.30}$$

13.6 Three-Dimensional Linear Constitutive Relations

The relations previously discussed for viscoelastic materials, based on mechanical models, can be generalized to stress–strain relations in three dimensions. This is similar to generalizing a simple spring to the Hookean solid. First, we separate the hydrostatic parts of stress and strain in the form

$$p = -\frac{1}{3}\sigma_{ii}, \quad \epsilon = \epsilon_{ii}, \tag{13.31}$$

where, for infinitesimal strain ϵ_{ij}, the quantity ϵ, as defined here, has the significance that it represents change in volume per unit volume.

For most materials, the assumption

$$p(t) = -K\epsilon(t), \tag{13.32}$$

where the bulk modulus K is taken as a constant, is satisfactory. If the response to hydrostatic pressure exhibits viscoelastic behavior, we may write

$$p(t) = -\int_0^t K(t-\tau)\frac{d\epsilon}{d\tau}\, d\tau, \tag{13.33}$$

where $K(t)$ may be called a hydrostatic stress relaxation function.

The deviatoric quantities

$$s_{ij} = \sigma_{ij} - \frac{1}{3}\sigma_{kk}\delta_{ij}, \quad e_{ij} = \epsilon_{ij} - \frac{1}{3}\epsilon_{kk}\delta_{ij} \tag{13.34}$$

may be related by use of the creep compliance and stress relaxation functions in the form

$$s_{ij} = \int_0^t G(t-\tau)\frac{de_{ij}}{d\tau}\,d\tau, \quad e_{ij} = \int_0^t J(t-\tau)\frac{ds_{ij}}{d\tau}\,d\tau. \tag{13.35}$$

The kernels K, G, and J are called hereditary kernels. These relations are applicable for isotropic, homogeneous, linear viscoelasticity. In terms of the Laplace transforms of quantities involved, we have

$$\bar{s}_{ij} = s\overline{G}\bar{e}_{ij}, \quad \bar{e}_{ij} = s\overline{J}\bar{s}_{ij}. \tag{13.36}$$

If we had a relation between s_{ij} and e_{ij} in the differential operator form,

$$\mathcal{P}(s)\bar{s}_{ij} = \mathcal{Q}(s)\bar{e}_{ij}, \tag{13.37}$$

then the stress relaxation and creep compliance functions G and J are given by

$$\overline{G} = \frac{1}{s}\frac{\overline{\mathcal{Q}}}{\overline{\mathcal{P}}}, \quad \overline{J} = \frac{1}{s}\frac{\overline{\mathcal{P}}}{\overline{\mathcal{Q}}}. \tag{13.38}$$

We may factor $\mathcal{P}$ and $\mathcal{Q}$ in the form

$$s\mathcal{P} = (s+1/\tau_{s1})^{m_1}(s+1/\tau_{s2})^{m_2}\cdots(s+1/\tau_{sp})^{m_p}, \quad m_1+m_2+\cdots+m_p = m+1, \tag{13.39}$$

$$s\mathcal{Q} = (s+1/\tau_{e1})^{n_1}(s+1/\tau_{e2})^{n_2}\cdots(s+1/\tau_{eq})^{n_q}, \quad n_1+n_2+\cdots+n_q = n+1. \tag{13.40}$$

Using these factored forms in Eqs. (13.38), we express G and J as partial fractions. Noting

$$\mathcal{L}[e^{-t/\tau}] = \int_0^\infty e^{-(s+1/\tau)t}\,dt = \frac{1}{s+1/\tau}, \quad \mathcal{L}[t^n e^{-t/\tau}] = \frac{n!}{(s+1/\tau)^{(n+1)}}, \tag{13.41}$$

we see multiple stress relaxation times τ_{sj}, $j = 1, 2, \ldots, p$, and strain relaxation times τ_{ei}, $i = 1, 2, \ldots, q$, while inverting the partial fraction expressions for G and J. If we encounter factors in which $1/\tau = 0$, that is, $1/s^i$ is a factor, its inverse will give a term like $t^{(i-1)}$. Such terms in J represent fluid-type motion without any decay with time. We have seen this behavior in Maxwell fluids. We have tacitly assumed all relaxation times obtained from our factoring $\overline{\mathcal{P}}$ and $\overline{\mathcal{Q}}$ are nonnegative. If any of them is negative (say in $\mathcal{Q}$), a slight force can set off an exponentially growing displacement, violating material stability. We need the existence of nonnegative factors as a constraint on the coefficients of the operational polynomials $\mathcal{P}$ and $\mathcal{Q}$.

13.7 Anisotropy

In the anisotropic case, we use the Voigt notation introduced in Chapter 11 to relate the deviatoric stress and strain in the form

$$\bar{s}_i = s\overline{G}_{ij}\bar{e}_j, \quad \bar{e}_i = s\overline{J}_{ij}\bar{s}_j, \quad i, j = 1, 2, \ldots, 6. \tag{13.42}$$

13.8 Biot's Theory

We consider the interesting results obtained by Biot (1955) in irreversible thermodynamics with applications to linear viscoelasticity. Although rapidly evolving thermodynamic state variables are not uncommon in many chemical and physical systems (examples can be found in chemical reactions and turbulent flows), Biot's theory concerns slowly evolving state variables close to their equilibrium values.

Consider a homogeneous thermodynamic system I, defined by n state variables q_i, which are measured from a reference state $q_i = 0$. These variables may represent displacements, shear angles, concentrations, temperature rises, etc. We associate a set of external forces P_i with q_i as conjugate variables. Let S_I represent the entropy of system I. As system I is not isolated, we create an isolated system by embedding system I in a second system, II, which is a large reservoir at temperature T_0; the combined system can be considered to be isolated. The total entropy of the combined system is

$$S = S_\mathrm{I} + S_\mathrm{II}. \tag{13.43}$$

When the two systems are brought into contact, let δQ_I and δQ_II be the heat added to the two systems. From the balance of energy,

$$\delta Q_\mathrm{I} = -\delta Q_\mathrm{II}. \tag{13.44}$$

For system I,

$$\delta Q_\mathrm{I} = dE_\mathrm{I} - P_i dq_i, \tag{13.45}$$

$$\delta Q_\mathrm{II} = -dE_\mathrm{I} + P_i dq_i. \tag{13.46}$$

The entropy change in the reservoir is then given by

$$dS_\mathrm{II} = \frac{1}{T_0}\delta Q_\mathrm{II} = \frac{1}{T_0}[-dE_\mathrm{I} + P_i dq_i]. \tag{13.47}$$

The total entropy change in the isolated system can be expressed as

$$dS = dS_\mathrm{I} + dS_2 = dS_\mathrm{I} + \frac{1}{T_0}[-dE_\mathrm{I} + P_i dq_i]. \tag{13.48}$$

According to the second law, $dS > 0$ when at least one $dq_i \neq 0$. As S_I and E_I are functions of the state variables, we have

$$X_i \equiv \frac{\partial S_\mathrm{I}}{\partial q_i} - \frac{1}{T_0}\left[\frac{\partial E_\mathrm{I}}{\partial q_i} + P_i\right] > 0, \tag{13.49}$$

where we have introduced the notation X_i associated with q_i for the entropy-producing force separate from P_i. Note that

$$dS = X_i dq_i, \quad \dot{S} = X_i \dot{q}_i, \tag{13.50}$$

where $\dot{q}_i$ is the flux associated with the generalized force X_i. We have seen this form of entropy production rate in Chapter 9.

We define an equilibrium state in which all external forces are zero, i.e., $P_i = 0$, and the total entropy is

$$S_{\text{eq}} = S_{\text{I}} - \frac{E_{\text{I}}}{T_0}. \tag{13.51}$$

At equilibrium the entropy of the system is a maximum, and for small values of q_i we may write

$$T_0 S_{\text{I}} - E_{\text{I}} = -\frac{1}{2} a_{ij} q_i q_j \equiv -U, \tag{13.52}$$

where a_{ij} is a symmetric, positive-definite matrix. We use the notation U for the quadratic form because of its similarity to the Helmholtz free energy. In terms of U, Eq. (13.48) can be expressed as

$$T_0 dS = -dU + P_i dq_i. \tag{13.53}$$

From the reference state $q_i = 0$, if we slowly increase the values of q_i in a quasi-static, reversible manner, at each equilibrium step we would have

$$T_0 \dot{S} = -\dot{U} + P_i \dot{q}_i = 0 \quad \text{or} \quad T_0 \frac{\partial S}{\partial q_i} = -\frac{\partial U}{\partial q_i} + P_i = 0. \tag{13.54}$$

In the irreversible case, we use Onsager's principle, and the entropy production rate is nonzero. That is,

$$T_0 \frac{\partial S}{\partial q_i} = b_{ij} \dot{q}_j, \tag{13.55}$$

where b_{ij} is a symmetric, positive-semidefinite (some state variables may not participate in entropy production) matrix. Equation (13.54) can be written as the evolution equations

$$\frac{\partial U}{\partial q_i} + b_{ij} \dot{q}_j = P_i \quad \text{or} \quad b_{ij} \dot{q}_j + a_{ij} q_j = P_i. \tag{13.56}$$

Biot has introduced a dissipation function D as

$$D = \frac{1}{2} b_{ij} \dot{q}_i \dot{q}_j, \tag{13.57}$$

which can be introduced in the evolution equation, Eq. (13.56), to get

$$\frac{\partial D}{\partial \dot{q}_i} + \frac{\partial U}{\partial q_i} = P_i, \tag{13.58}$$

which has a resemblance to the Lagrange equations of dynamics. Also,

$$T_0 \frac{\partial S}{\partial t} = T_0 \frac{\partial S}{\partial q_i} \dot{q}_i = b_{ij} \dot{q}_i \dot{q}_j = 2D. \tag{13.59}$$

In terms of the dissipation function D, the entropy-producing forces X_i in Eq. (13.50) can be written as

$$\frac{\partial D}{\partial \dot{q}_i} = X_i, \tag{13.60}$$

and the power input as

$$X_i \dot{q}_i = 2D. \tag{13.61}$$

13.8.1 Minimum Entropy Production Rate

For a given power input $X_i \dot{q}_i$, one may ask the question: Of all possible sets of values for $\dot{q}_i$, which set makes the entropy production rate a minimum?

Using Eq. (13.59), we want to minimize D subject to the constraint $X_i \dot{q}_i = k$. With the help of a Lagrange multiplier λ, we define a modified function

$$D^* = D - \lambda(X_i \dot{q}_i - k), \tag{13.62}$$

and minimize it to get

$$\frac{\partial D}{\partial \dot{q}_i} = \lambda X_i, \tag{13.63}$$

as the equations for the general directions of $\dot{q}_i$. If we multiply these by $\dot{q}_i$ (and sum on i), we find

$$2D_{\text{min}} = \lambda k, \tag{13.64}$$

which simplifies to $\lambda = 1$ when Eq. (13.61) is used. Thus, our evolution equation, Eq. (13.60), corresponds to minimum entropy production rate for a given power input.

Biot (1955) and Fung (1965) are highly recommended for further reading on this topic. These suggested readings also include a method for eliminating hidden variables q_i (which do not have any associated generalized external forces P_i).

13.9 Creep in Metals

In uniaxial experiments conducted at high temperatures, it is seen that the strain rate depends on the stress in a nonlinear fashion. Generally, the strain versus time plot can be divided into three domains. These are called the primary creep, secondary creep, and tertiary creep domains. A sketch of the three domains is shown in Fig. 13.8. In the case of primary creep, the following constitutive law is often used to describe the creep behavior:

$$\dot{e} = A s^n t^m, \tag{13.65}$$

where A, n, and m are constants. For a multiaxial state of stress, using deviatoric quantities, we can have

$$\dot{e}_{ij} = A I_{s2}^{n-1} t^m s_{ij}, \tag{13.66}$$

where I_{s2} is the second invariant of the deviatoric stress s_{ij}. The assumption that the material is incompressible is usually applied.

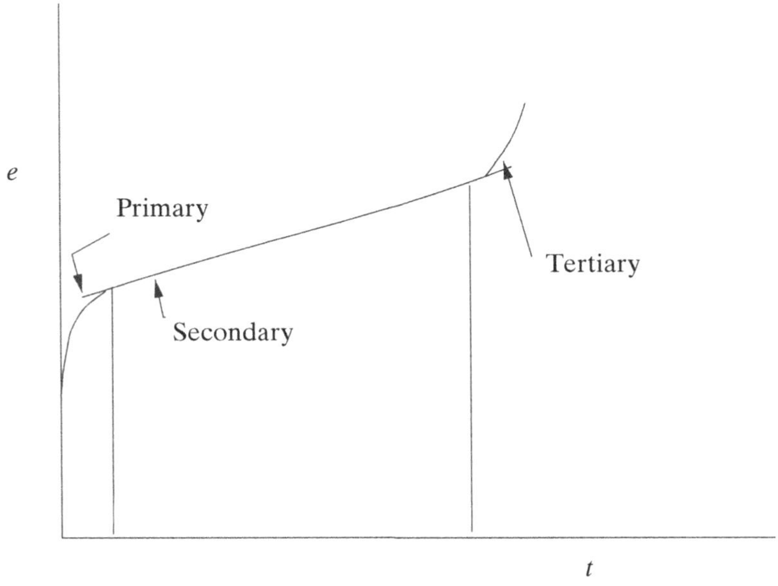

Figure 13.8. Three domains of creep behavior.

For the secondary creep, the preceding law with $m = 0$ is used. This domain is also known as the domain of stationary or steady-state creep. The tertiary creep behavior is seen just before the specimen fails.

Another aspect of creep in materials involves differing strain rate responses to tensile and compressive loads. That is, the creep constants, A, n, and m for compression, are different from those for tension.

13.10 Nonlinear Theories of Viscoelasticity

The infinitesimal strains used in the preceding sections are not applicable when large deformations are involved. Also, the material response itself may depend on stresses and strains in a nonlinear fashion. The elementary constitutive relations previously discussed have been extended, in a number of ways, for nonlinear viscoelastic solids and fluids. An extensive classification of viscoelastic materials can be found in the book by Eringen (1962). A brief introduction to this topic is subsequently given.

We begin by introducing an intermediate time τ between the initial time $\tau = 0$ and the current time $\tau = t$. We also have an intermediate location χ corresponding to the intermediate time. Then

$$\boldsymbol{X} = \boldsymbol{\chi}(x, 0), \quad \boldsymbol{\chi} = \boldsymbol{\chi}(x, \tau), \quad \boldsymbol{x} = \boldsymbol{\chi}(x, t). \tag{13.67}$$

Extending the Greek indices for the material variables and Latin indices for the current variables, we use a, b, c, etc., for the intermediate variables:

$$\boldsymbol{\chi} = \chi_a \mathbf{e}_a. \tag{13.68}$$

The Cauchy deformation gradient $\boldsymbol{f}$ and the Green's deformation gradient $\boldsymbol{F}$ are defined for the intermediate time τ with the components

$$f_{ia} = \frac{\partial \chi_a}{\partial x_i}, \quad F_{ai} = \frac{\partial x_i}{\partial \chi_a}, \tag{13.69}$$

and the corresponding deformation tensors,

$$g_{ij} = f_{ia} f_{ja}, \quad G_{ab} = F_{ai} F_{bi}. \tag{13.70}$$

The Almansi and Lagrange strains are given by

$$e_{ij} = \frac{1}{2}[\delta_{ij} - g_{ij}], \quad E_{ab} = \frac{1}{2}[G_{ab} - \delta_{ab}]. \tag{13.71}$$

As can be seen from the preceding definitions, what we have is an Eulerian description with the intermediate position in the role of the initial position $\boldsymbol{X}$ we had used before. Next, we introduce higher-order intermediate strain rate and stress rate variables. These are generalized versions of the deformation rate and Jaumann stress rates. We define

$$\left(\frac{D}{D\tau}\right)^M (d\chi_a d\chi_a) = A_{ab}^{(M)} d\chi_a d\chi_b, \tag{13.72}$$

where $A_{ab}^{(M)}$ is called the Rivlin–Ericksen tensor of the order of M. When $M = 1$ we get the intermediate deformation rate tensor. Starting with

$$s_{ab}^{(0)} = \sigma_{ab}, \tag{13.73}$$

we define higher-order objective rates as

$$s_{ab}^{(N)} = \dot{s}_{ab}^{(N-1)} + s_{ac}^{(N-1)} \partial_b v_c + s_{bc}^{(N-1)} \partial_a v_c, \tag{13.74}$$

where v_c is the velocity component at time τ. It can be shown that these stress rates are objective. With this, the general constitutive relations for viscoelastic materials can be expressed as

$$\mathop{\mathcal{F}_{cd}}_{\tau=0}^{\tau=t} \left[\partial_\alpha \chi_a, A_{ab}^{(1)}(\tau), A_{ab}^{(2)}(\tau), \ldots, A_{ab}^{(M)}(\tau), s_{ab}^{(0)}(\tau), s_{ab}^{(1)}(\tau), \ldots, s_{ab}^{(N)}(\tau)\right] = 0. \tag{13.75}$$

From this, we can obtain various subclasses of materials, including Stokes fluid and Cauchy elastic materials, by specifying the values of M and N and the functional forms.

13.11 K-BKZ Model for Viscoelastic Fluids

Among the theories of viscoelastic fluids, the K-BKZ theory has found widespread applications in modeling polymer fluids. This model is based on two independent papers by Kaye (1962) and Bernstein, Kearsley, and Zapas (1963). Our subsequent discussion is based on a review of the theory by Tanner (1988).

We begin with the hereditary integral representation of the shear stress in a one-dimensional pure shear experiment, as

$$s(t) = \int_{-\infty}^{t} G(t - \tau)\dot{E} d\tau, \tag{13.76}$$

where E is the shear strain. Integrating by parts and assuming $G(0) = 0$ and $G(-\infty) = 0$, we have

$$s(t) = \int_{-\infty}^{t} \dot{G}(t - \tau) E d\tau. \tag{13.77}$$

Nonlinear dependence on E is then introduced by replacing $\dot{G}$ with $\partial U/\partial E$, where $U = U[E(\tau), t - \tau]$. The original K-BKZ model assumes incompressibility with an associated hydrostatic pressure. The 3D version of the model is

$$s_{ij}(t) = \int_{-\infty}^{t} F_{ai} \frac{\partial U}{\partial E_{ab}} F_{bj} d\tau. \tag{13.78}$$

The invariance requirements restrict, for isotropic, incompressible fluids, the form of the **memory kernel** U to

$$U = U(I_{E1}, I_{E2}). \tag{13.79}$$

As we have done in the case of finite elasticity, if U is assumed to be a function of the invariants of the deformation gradients G_{ab},

$$s_{ij}(t) = \int_{-\infty}^{t} F_{ai} \frac{\partial U}{\partial G_{ab}} F_{bj} d\tau. \tag{13.80}$$

Tanner (1988) gives a number of special functional forms for U and their successes and failures when applied to various polymeric fluids.

The original isothermal theory was extended by Bernstein, Kearsley, and Zapas (1964) in a subsequent paper.

SUGGESTED READING

Bernstein, B., Kearsley, E. A., and Zapas, L. J. (1963). A study of stress relaxation with finite strain, *Trans. Soc. Rheol.*, **7**, 391–410.

Bernstein, B., Kearsley, E. A., and Zapas, L. J. (1964). Thermodynamics of perfect elastic fluids, *J. Res. Natl. Bur. Stand.*, **68B**, No. 3, 103–113.

Biot, M. A. (1955). Variational principles in irreversible thermodynamics with applications to viscoelasticity, *Phys. Rev.*, **97**, 1463–1469.

Eringen, A. C. (1962). *Nonlinear Theory of Continuous Media*, McGraw-Hill.

Fung, Y. C. (1965). *Foundations of Solid Mechanics*, Prentice-Hall.

Kaye, A. (1962). Non-Newtonian flow in incompressible fluids, Part I: A general rheological equation of state, Part II: Some problems in steady flow, Note No. 134, College of Aeronautics, Cranfield, UK.

Rabotnov, Y. N. (1969). *Creep Problems in Structural Members*, North-Holland.

Tanner, R. I. (1988). From A to (BK)Z in constitutive relations, *J. Rheol.*, **32**, 673–702.

EXERCISES

13.1. Consider a bar of cross-sectional area A, length L, and density ρ. One end of the bar, $x = L$, is fixed to a wall and the other end, $x = 0$, is subjected to an oscillating force, $F(t) = F_0 e^{i\omega t}$. Assuming the material is a Kelvin–Voigt solid satisfying

$$\sigma = E\epsilon + \mu\dot{\epsilon},$$

obtain the displacement under the load, neglecting changes in the cross-sectional area and the inertia forces. Reduce the displacement to its steady-state form. Define the so-called complex modulus E_c, which is a complex number involving E, μ, and ω, by comparing the solution with the elastic solution.

13.2. Assume the bar in the preceding exercise extends from $x = 0$ to ∞. If a stress σ_0 is suddenly applied at $x = 0$, obtain the displacement $u(x, t)$ as a real integral. Neglect the change in the cross-sectional area.

13.3. Calculate the response of a standard linear solid when

$$P = P_0 t H(t).$$

13.4. A mechanical model is shown in Fig. 13.9. Obtain the viscoelastic model for this setup.

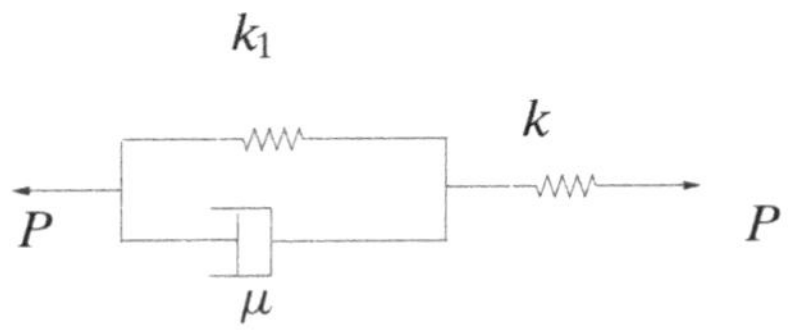

Figure 13.9. A mechanical model for a viscoelastic solid.

13.5. The time history of the applied load on a viscoelastic bar is shown in Fig. 13.10. Using the superposition integral, obtain the displacement history if the material is (a) a Maxwell fluid, and (b) a standard linear solid.

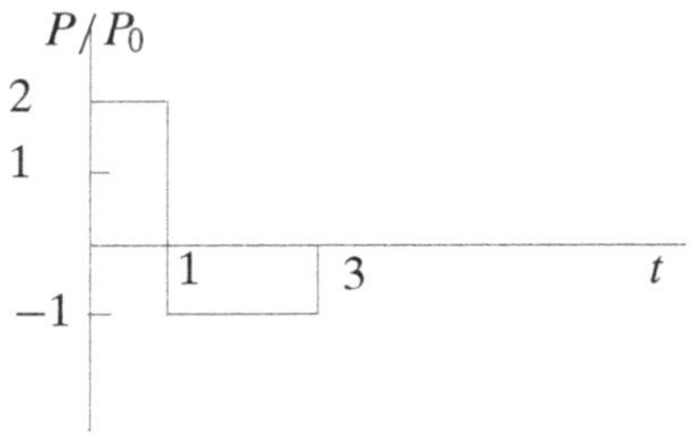

Figure 13.10. History of applied load.

13.6. For a 3D viscoelastic body, write the field equations and boundary and initial conditions under the assumption of infinitesimal strains. Using the Laplace transform, convert these equations into equations in the transformed variables. Compare with the equations of linear elasticity and establish the **correspondence principle**. Use Eqs. (13.33) and (13.35) as constitutive laws. Comment on the boundary conditions for which the correspondence principle fails.

13.7. A cube of side length a is made of a Maxwell material. It is placed on a frictionless surface at time $t = 0$. Because of the weight of the material, there is a compressive stress at any cross-section. Assuming a specific gravity of γ, compute the compressive stress at a height x from the bottom of the cube. Obtain the height of the cube at any time t. Neglect changes in the cross-sectional area.

13.8. A Maxwell fluid with the constitutive relation

$$\frac{\partial u}{\partial y} = \frac{1}{G}\dot{\tau} + \frac{1}{\mu}\tau$$

occupies the region $0 < y < h$, $-\infty < x, z < \infty$. At the boundary $y = 0$, the velocity $u = 0$, and at $y = h$, $u = Ue^{i\omega t}$. Neglecting any body forces, obtain the velocity distribution $u(y, t)$.

14 Plasticity

The time-independent, permanent deformation in metals beyond the elastic limit is described by the term **plasticity**. As shown in Fig. 14.1, the elastic part of the uniaxial stress–strain curve OA is reversible. When we unload from any point beyond point A, corresponding to the zero-stress state, there is a permanent deformation in the specimen. Several theories describing the plastic behavior of metals have been proposed, but none is totally satisfactory. Ideally, the stress σ, separating the elastic part OA from the inelastic part, is called the **yield stress** σ_0 of the material. However, it is difficult to obtain this special point accurately from experiments, and it often depends on the history of the loading the specimen has undergone.

When a specimen is unloaded from a point C to zero stress and reloaded, permanent deformation usually begins at a stress level σ_C. In other words, the yield stress is higher after the specimen has undergone a certain amount of permanent deformation. This is known as **work hardening** or **strain hardening**.

Another feature of metal deformation is that, from a strained state, beyond the yield point, reversal of loading causes the compressive yield stress to be different from the tensile yield stress. It is usually lower in magnitude than the tensile yield stress. This effect is called the **Bauschinger effect**.

Before we consider modern theories of plasticity that attempt to describe the multiaxial stress–strain behavior, let us review some idealized classical theories in the next few sections of this chapter.

14.1 Idealized Theories

14.1.1 Rigid Perfectly Plastic Material

In a uniaxial stress state the material is assumed to be rigid up to a value of stress $\sigma = \sigma_0$ (see Fig. 14.2). At σ_0, the material flows. Because of this limiting upper bound for the stress, it is to be kept in mind that all plasticity experiments (including thought experiments) are done with displacement control. The term **perfectly plastic** refers to the absence of work hardening. In applications such as metal forming, in which the elastic strains are small, this idealization is still used to a great extent.

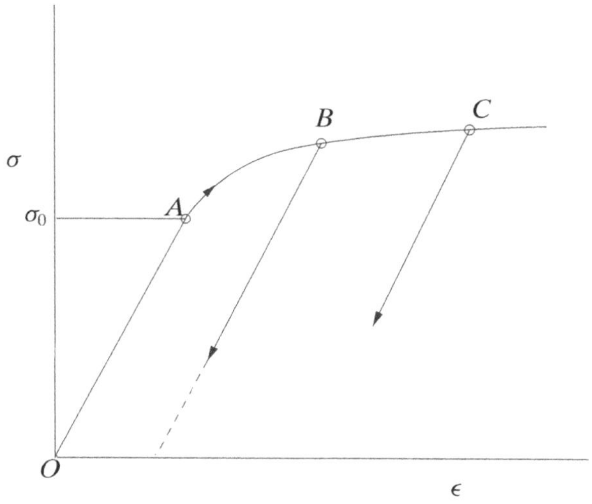

Figure 14.1. Stress–strain relation for metals in uniaxial tension.

14.1.2 Elastic Perfectly Plastic Material

When $\sigma < \sigma_0$ the material obeys Hooke's law, and when $\sigma \geq \sigma_0$ the material flows (see Fig. 14.3). In problems concerned with residual stresses the elastic effects are significant and this idealization is often used.

14.1.3 Elastic Linearly Hardening Material

The uniaxial stress–strain behavior previously described can be extended to work-hardening models through bilinear relations, as shown in Fig. 14.4.

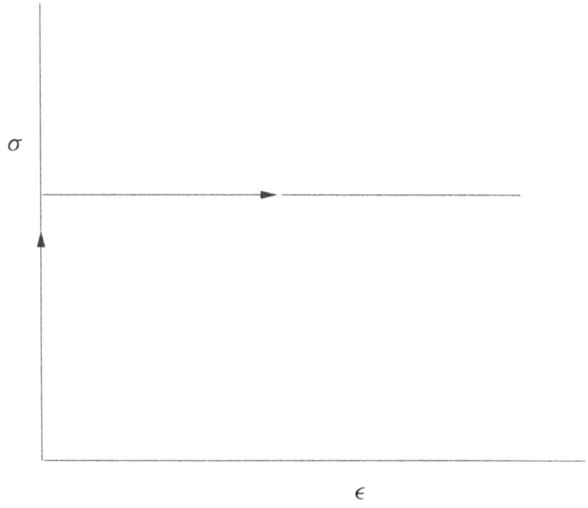

Figure 14.2. Rigid perfectly plastic model.

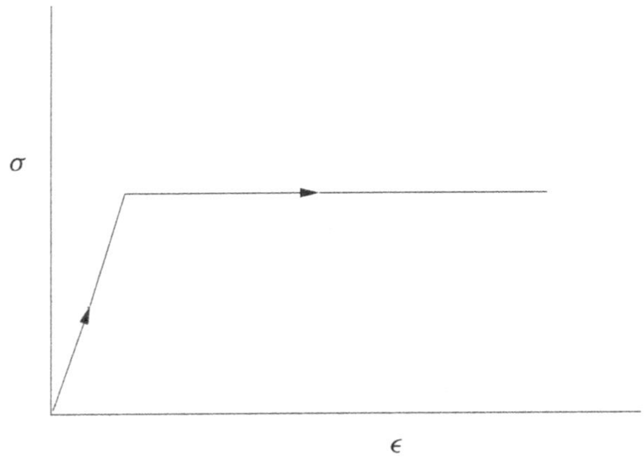

Figure 14.3. Elastic perfectly plastic model.

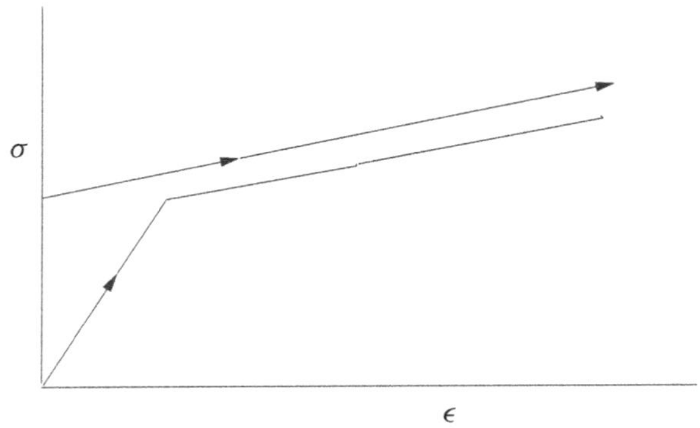

Figure 14.4. Linear work-hardening models.

Plastic flow problems can be generally classified as **contained plastic flow** and **uncontained plastic flow**. These two cases are illustrated by Fig. 14.5, by considering the tip of a crack in a plate.

In uncontained plastic flow, the work hardening is important and the elastic effects may be negligible. However, in contained plastic flow the elastic effects are significant.

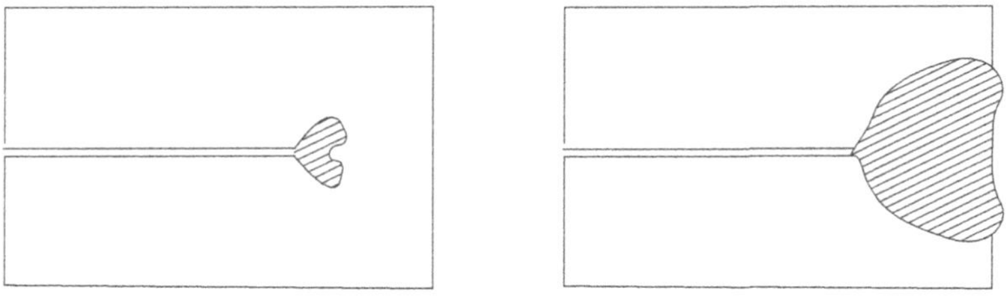

Figure 14.5. Contained and uncontained plastic domains.

14.2 Three-Dimensional Theories

When we attempt to extend the uniaxial behavior to the three-dimensional (3D) stress state, an important factor to observe is that the material behavior under hydrostatic stress is significantly different from that under pure shear. The onset of permanent deformation is governed by the state of shear stress in most polycrystalline, metallic materials. Spitzig and Richmond (1984) provide Aluminum 1100 as an example of pressure-sensitive plastic yield behavior. In this chapter we exclude pressure-dependent yielding. We classify the elastic behavior of materials as compressible or incompressible on the basis of the stress–strain behavior under hydrostatic pressure. The relation for the hydrostatic pressure can be written as

$$-p = K\epsilon, \qquad \text{compressible}, \tag{14.1}$$

$$p = \bar{p}, \qquad \text{incompressible}, \tag{14.2}$$

where

$$p = -\frac{1}{3}\sigma_{ii}, \qquad \epsilon = \epsilon_{ii}. \tag{14.3}$$

We denote the deviatoric stress by

$$s_{ij} = \sigma_{ij} - \frac{1}{3}\sigma_{kk}\delta_{ij}, \tag{14.4}$$

and the deviatoric strain by

$$e_{ij} = \epsilon_{ij} - \frac{1}{3}\epsilon_{kk}\delta_{ij}. \tag{14.5}$$

To distinguish the range of s_{ij}, in which the material behaves elastically, we need a yield condition. If, under uniaxial tension, the yield stress is σ_0, the corresponding shear stress τ_0 is given by

$$\tau_0 = \sigma_0/2, \tag{14.6}$$

and we would like to extend this as a condition on the stress state in the 3D case. This constraint should involve only the deviatoric stresses. For most metals, experiments indicate that the following two yield conditions may be used as approximations:

$$f(\boldsymbol{s}) = \frac{1}{2}\max|(\sigma_1 - \sigma_2), (\sigma_2 - \sigma_3), (\sigma_3 - \sigma_1)| - k_1 = 0, \tag{14.7}$$

$$f(\boldsymbol{s}) = \frac{1}{2}s_{ij}s_{ij} - k_2 = 0. \tag{14.8}$$

These two conditions are known as the Tresca condition and the von Mises condition, respectively. The Tresca condition states that plastic flow commences when the maximum shear stress reaches a specified value. The von Mises condition is based on the distortion energy density stored in the material. Distortion energy is the difference between the total strain energy and the work done by the hydrostatic pressure.

To examine these conditions closely, it is advantageous to use the stress tensor referred to as the principal axes:

$$\boldsymbol{\sigma} = \begin{bmatrix} \sigma_1 & 0 & 0 \\ 0 & \sigma_2 & 0 \\ 0 & 0 & \sigma_3 \end{bmatrix}, \tag{14.9}$$

where σ_i are the principal stresses and $\sigma_1 \geq \sigma_2 \geq \sigma_3$.

Here, the hydrostatic stress is given by

$$p = -\frac{1}{3}(\sigma_1 + \sigma_2 + \sigma_3). \tag{14.10}$$

Then,

$$\boldsymbol{s} = \frac{1}{3} \begin{bmatrix} 2\sigma_1 - \sigma_2 - \sigma_3 & 0 & 0 \\ 0 & 2\sigma_2 - \sigma_3 - \sigma_1 & 0 \\ 0 & 0 & 2\sigma_3 - \sigma_1 - \sigma_2 \end{bmatrix}, \tag{14.11}$$

and the Tresca criterion reads

$$\sigma_1 - \sigma_3 - 2k_1 = 0, \tag{14.12}$$

and the von Mises criterion reads

$$J_{\sigma 2} = I_{s2} = \frac{1}{2} s_{ij} s_{ij} = \frac{1}{6}[(\sigma_1 - \sigma_2)^2 + (\sigma_2 - \sigma_3)^2 + (\sigma_3 - \sigma_1)^2] - k_2 = 0, \tag{14.13}$$

where we have introduced the notation $J_{\sigma i} = I_{si}$ relating the invariants of the deviatoric stress and the full stress tensor. Noting that, when the material yields under uniaxial tension, the stress state is

$$\sigma_1 = \sigma_0, \quad \sigma_2 = 0, \quad \sigma_3 = 0, \tag{14.14}$$

we find the constants

$$k_1 = \sigma_0/2, \qquad k_2 = \sigma_0^2/3. \tag{14.15}$$

In the stress space, with the axes σ_1, σ_2, and σ_3, we have

$$\sigma_1 + \sigma_2 + \sigma_3 = -3p, \tag{14.16}$$

representing planes with unit normal

$$\boldsymbol{n} = \frac{1}{\sqrt{3}}(1, 1, 1)^{\mathrm{T}}. \tag{14.17}$$

These planes are called the π planes. Because the yield condition does not depend on any of these parallel planes, it represents a surface normal to these planes. In other words, the yield surface is an orthogonal cylinder with its axis inclined at an angle θ with respect to the three principal directions, where $\cos\theta = 1/\sqrt{3}$, as shown

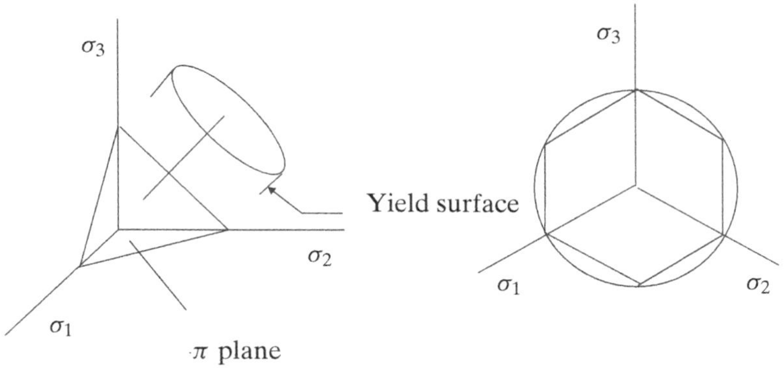

Figure 14.6. Yield surfaces in the stress space.

in Fig. 14.6. A stress state $\boldsymbol{\sigma}$, which is a vector in the stress space with σ_1, σ_2, and σ_3 as coordinates, can be resolved as

$$\boldsymbol{\sigma} = -p\boldsymbol{n} + \boldsymbol{\tau}, \qquad \boldsymbol{\tau} \cdot \mathbf{n} = 0. \tag{14.18}$$

The three components of $\boldsymbol{\tau}$ are

$$\tau_1 = \frac{1}{3}(2\sigma_1 - \sigma_2 - \sigma_3), \quad \tau_2 = \frac{1}{3}(2\sigma_2 - \sigma_3 - \sigma_1), \quad \tau_3 = -(\tau_1 + \tau_2). \tag{14.19}$$

The yield surface is a closed curve in the two-dimensional (2D) τ_1, τ_2 space. Instead of τ_1 and τ_2, we may use the more familiar quantities $\sigma_1 - \sigma_3$ and $\sigma_2 - \sigma_3$ for plotting the yield curve.

The Tresca condition becomes an elongated hexagon in our plot and the von Mises condition an elongated ellipse. In the π plane, these figures will lose their elongations and become a perfect hexagon and a circle.

Combined torsion and tension of thin-walled tubes is often used to generate a multiaxial stress state for determining whether the Tresca condition or the von Mises condition is a better approximation for the yield condition for a particular material.

With tensile stress σ and the shear stress tensor τ, the stress tensor can be written as

$$\boldsymbol{\sigma} = \begin{bmatrix} \sigma & \tau & 0 \\ \tau & 0 & 0 \\ 0 & 0 & 0 \end{bmatrix}, \tag{14.20}$$

with the principal stresses

$$\sigma_{1,3} = \frac{1}{2}[\sigma \pm \sqrt{\sigma^2 + 4\tau^2}], \quad \sigma_2 = 0 \tag{14.21}$$

and the maximum shear

$$\tau_{\max} = \frac{1}{2}\sqrt{\sigma^2 + 4\tau^2}. \tag{14.22}$$

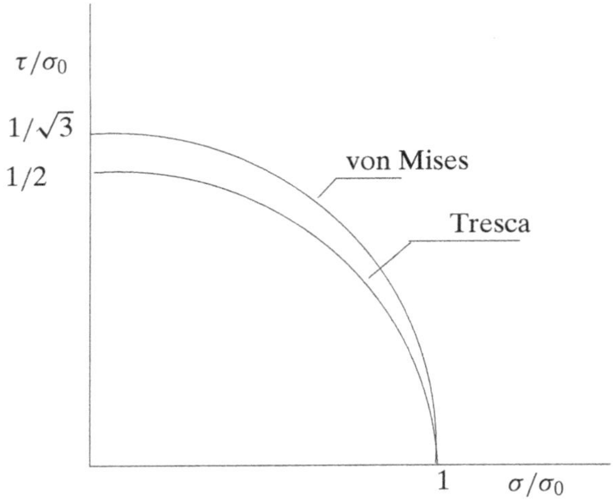

Figure 14.7. Yield curves in the tension–torsion of thin-walled tubes.

The value of J_2 is obtained as

$$J_2 = \frac{1}{3}[\sigma^2 + 3\tau^2]. \tag{14.23}$$

A plot of Tresca and von Mises Conditions for the tension-torsion experiment is shown in Fig. 14.7.

Once a yield condition has been established for a work-hardening material, σ_0 will change with strain history. In materials such as soil and other granular materials, the yield condition is known to depend on the hydrostatic pressure p. The higher the value of p, the larger will be the yield stress.

In many applications, plane stress or plane strain approximations are satisfactory and it is worthwhile to examine the yield conditions for these cases. To do this, first we need to consider the relation between plastic strain and the stress state.

14.3 Postyield Behavior

Uniaxial experiments show that, under controlled strain rates, the stress increases to the yield stress and a further increase in strain does not increase the stress substantially. The rate of loading has negligible effect on the stress at room temperature. Thus plastic flow is independent of time, unlike fluid flow, in which the shear stress is proportional to the strain rate. We have to keep in mind that the "loading" is applied through controlled displacements on the boundary. Two distinct constitutive relations are in use, depending on the rigid perfectly plastic or the elastic perfectly plastic approximation. These are known as the flow rules.

14.3.1 Levy–Mises Flow Rule

Following the rigid perfectly plastic approximation, the elastic strains are neglected. Furthermore, the material is assumed to be incompressible. Thus,

$$\epsilon_{ij} = e_{ij} = e^p_{ij}, \tag{14.24}$$

where e_{ij} represents the inelastic strain. The constitutive relation is given by

$$de_{ij} = \sqrt{3}[\frac{1}{2}de_{kl}de_{kl}]^{1/2}\frac{s_{ij}}{\sigma_0}. \tag{14.25}$$

Using

$$d\lambda = \sqrt{3J_2(\boldsymbol{de})}, \tag{14.26}$$

we have

$$de_{ij} = d\lambda s_{ij}/\sigma_0. \tag{14.27}$$

From dissipation considerations, it can be shown that the von Mises criterion is the natural criterion to use with the Levy–Mises flow rule.

For plane strain deformations, $de_{33} = 0$ and $\sigma_{33} = (\sigma_{11} + \sigma_{22})/2$. Using this and the time derivatives

$$de_{ij} = \frac{de_{ij}}{dt}dt, \tag{14.28}$$

we find, for both plane stress and plane strain,

$$\frac{\dot{e}_{11}}{\sigma_{11} - \sigma_{22}} = \frac{\dot{e}_{22}}{\sigma_{22} - \sigma_{11}} = \frac{\dot{e}_{12}}{\sigma_{12}}. \tag{14.29}$$

14.3.2 Prandtl–Reuss Flow Rule

We begin with a decomposition of the strain increment as

$$d\epsilon_{ij} = d\epsilon^e_{ij} + d\epsilon^p_{ij}, \tag{14.30}$$

where $d\epsilon^e$ indicates the elastic component of the strain. We assume plastic incompressibility, that is,

$$d\epsilon^p_{ii} = 0 \quad \text{or} \quad d\epsilon^p_{ij} = de^p_{ij}. \tag{14.31}$$

The flow rule states that

$$de^p_{ij} = d\lambda s_{ij}/\sigma_0, \tag{14.32}$$

where $d\lambda$ is a plastic strain increment-dependent scalar. In contrast with the Levy–Mises rule, in which the total strain increment is used to define $d\lambda$, the Prandtl–Reuss flow rule has

$$d\lambda = \sqrt{3J_2(de^p)}. \tag{14.33}$$

The elastic strain increment is given by

$$d\epsilon^e_{ij} = ds_{ij}/2G, \qquad \epsilon^e = -p/K, \tag{14.34}$$

where K is the bulk modulus given by $K = E/3(1 - 2\nu)$.

14.4 General Yield Condition and Plastic Work

The yield condition can be represented by a hypersurface in the six-dimensional stress space σ_{ij}. Because the hydrostatic pressure has no effect on yielding, this hypersurface is not closed. If we use the five-dimensional space spanned by s_{ij}, we have a closed surface. The material behavior is elastic if the current stress state is inside this surface or if unloading takes place from the yield surface.

We insert a parameter k, representing the variable yield stress in the case of hardening plasticity, to express the yield condition as

$$f(\boldsymbol{s}; k) = 0. \tag{14.35}$$

The parameter k is considered to be an internal variable that depends on the history of the plastic strains, and an experimenter has no direct means to alter it. We have elastic constitutive relations if

$$f(\boldsymbol{s}; k) < 0 \quad \text{or} \quad f(\boldsymbol{s}; k) = 0, \quad \nabla f \cdot d\boldsymbol{s} < 0, \tag{14.36}$$

and plastic flow if

$$f(\boldsymbol{s}; k) = 0 \quad \text{and} \quad \nabla f \cdot d\boldsymbol{s} \geq 0. \tag{14.37}$$

This stress space description of the onset of plastic flow has to be considered in the context of strain-controlled experiments for clarity. If there is no hardening, the stress state is inside the yield surface or on it (see Fig. 14.8).

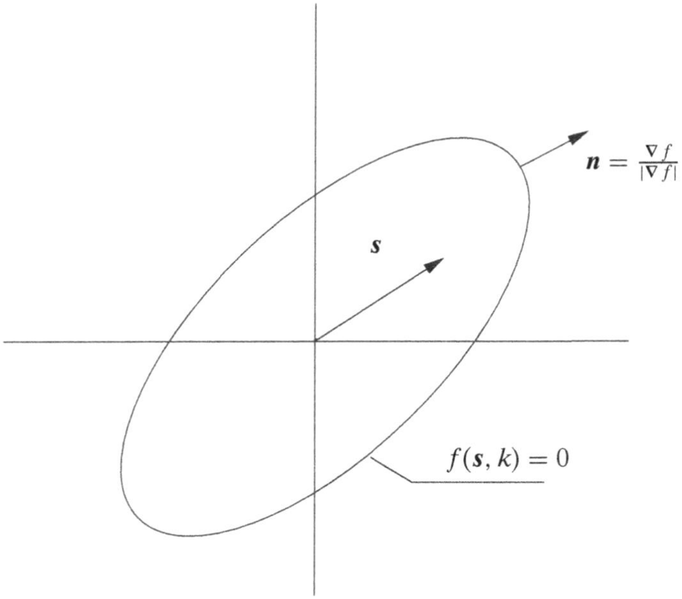

Figure 14.8. Yield surface in the deviatoric stress space.

14.4.1 Plane Stress and Plane Strain

For plane stress, the principal stress $\sigma_3 = 0$ and the von Mises condition becomes

$$\sigma_1^2 - \sigma_1\sigma_2 + \sigma_2^2 = \sigma_0^2. \tag{14.38}$$

In terms of a mean stress σ and a shear stress τ, defined by

$$\sigma \equiv \frac{\sigma_1 + \sigma_2}{2}, \quad \tau = \frac{\sigma_1 - \sigma_2}{2}, \tag{14.39}$$

where we assume $\sigma_1 > \sigma_2$, we find

$$\tau = \sqrt{(\sigma_0^2 - 2\sigma^2)/6}. \tag{14.40}$$

For plane strain, the plastic strain has to be confined to a plane normal to the x_3 axis and we get

$$\tau = \frac{\sigma_0}{2}. \tag{14.41}$$

The Tresca condition for plane strain gives

$$\tau = \frac{\sigma_0}{2}, \tag{14.42}$$

the same as the von Mises condition. However, for plane stress, we have different possibilities depending on the edge of the hexagon where the state of stress at yield happens to be.

14.4.2 Rigid Plasticity and Slip-Line Field

In many applications involving metal processing, the plastic strains are orders of magnitude larger than elastic strains and, if residual stresses are not of concern, the rigid plasticity approximation can be used. In the regions where metal flow is taking place, the maximum shear $\tau = \sigma_0/2$ under the plane strain approximation. As shown in Mohr's circle in Fig. 14.9, the stresses can be written as

$$\begin{aligned} \sigma_{11} &= \sigma + \tau \sin 2\theta, \\ \sigma_{22} &= \sigma - \tau \sin 2\theta, \\ \sigma_{12} &= -\tau \cos 2\theta, \end{aligned} \tag{14.43}$$

where

$$\sigma = \frac{1}{2}(\sigma_1 + \sigma_2), \quad \tau = \frac{1}{2}|\sigma_1 - \sigma_2| = \frac{1}{2}\sigma_0. \tag{14.44}$$

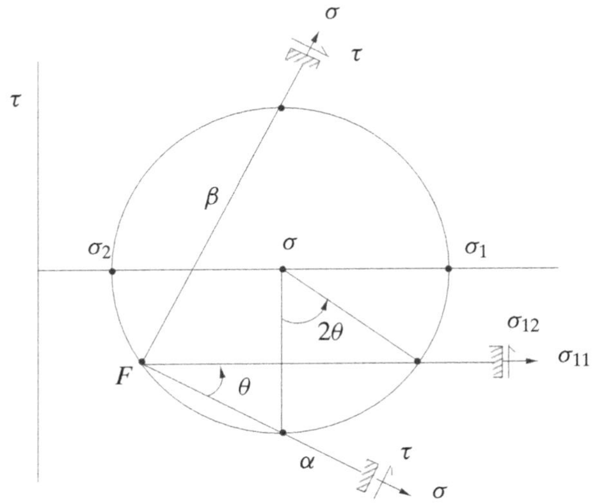

Figure 14.9. Mohr's circle and slip lines.

Using these in the equilibrium equations,

$$\sigma_{11,1} + \sigma_{12,2} = 0, \quad \sigma_{12,1} + \sigma_{22,2} = 0, \tag{14.45}$$

we find

$$\sigma_{,1} + \sigma_0[\cos 2\theta\, \theta_{,1} + \sin 2\theta\, \theta_{,2}] = 0, \tag{14.46}$$

$$\sigma_{,2} + \sigma_0[\sin 2\theta\, \theta_{,1} - \cos 2\theta\, \theta_{,2}] = 0. \tag{14.47}$$

These are two nonlinear equations for the mean stress σ and the angle of the maximum shear plane(s) θ. Following the method of characteristics introduced in Chapter 12, we define a characteristic variable α through

$$d\alpha = dx_1 - c\,dx_2 \quad \text{or} \quad \partial_2 = -c\partial_1. \tag{14.48}$$

As we will see, c has two values, and the corresponding two characteristic curves are denoted as α and β lines. The equilibrium equations give

$$\begin{bmatrix} 1 & \sigma_0(\cos 2\theta - c \sin 2\theta) \\ -c & \sigma_0(\sin 2\theta + c\cos 2\theta) \end{bmatrix} \begin{Bmatrix} \sigma_{,1} \\ \theta_{,1} \end{Bmatrix} = \begin{Bmatrix} 0 \\ 0 \end{Bmatrix}. \tag{14.49}$$

The characteristic equation of this eigenvalue problem is

$$c^2 - 2c\frac{\cos 2\theta}{\sin 2\theta} - 1 = 0, \tag{14.50}$$

which has two real solutions,

$$c = -\tan\theta, \quad c = \cot\theta. \tag{14.51}$$

These orthogonal directions are shown in Fig. 14.9. Using the eigenvectors corresponding to the two values of c, we obtain the Riemann invariants,

$$\frac{\sigma}{\sigma_0} + \theta = C_\beta, \quad \text{along the } \alpha \text{ line:} \quad \frac{dx_2}{dx_1} = -\tan\theta, \tag{14.52}$$

$$\frac{\sigma}{\sigma_0} - \theta = C_\alpha, \quad \text{along the } \beta \text{ line:} \quad \frac{dx_2}{dx_1} = \cot\theta. \tag{14.53}$$

The velocity gradients are obtained from the flow rule of Eq. (14.29) as

$$\frac{v_{1,1}}{\sigma_0 \sin 2\theta} = -\frac{v_{2,2}}{\sigma_0 \sin 2\theta} = -\frac{v_{1,2} + v_{2,1}}{2\sigma_0 \cos 2\theta}. \tag{14.54}$$

These relations can also be written as

$$v_{1,1} + v_{2,2} = 0, \quad v_{1,2} + v_{2,1} - 2v_{2,2} \cot 2\theta = 0. \tag{14.55}$$

Introducing the characteristic curves, with

$$\partial_2 = -c\partial_1, \tag{14.56}$$

we get

$$\begin{bmatrix} 1 & -c \\ -c & 1 + 2c\cot 2\theta \end{bmatrix} \begin{Bmatrix} v_{1,1} \\ v_{2,1} \end{Bmatrix} = \begin{Bmatrix} 0 \\ 0 \end{Bmatrix}. \tag{14.57}$$

The characteristic equation of this eigenvalue problem is the same as the one for stresses,

$$c^2 - 2c\frac{\cos 2\theta}{\sin 2\theta} - 1 = 0, \tag{14.58}$$

with the two real solutions, $c = \cot\theta$ and $c = -\tan\theta$.

For $c = \cot\theta$, we have

$$v_{1,1} = \cot\theta\, v_{2,1}, \tag{14.59}$$

and for $c = -\tan\theta$,

$$v_{1,1} = -\tan\theta\, v_{2,1}. \tag{14.60}$$

These are the incompressibility conditions expressed along the characteristics. If α and β are the curvilinear coordinates along the two characteristics, the flow rule shows that, along the maximum shear directions,

$$\sigma_{\alpha\alpha} = \sigma_{\beta\beta} = \sigma, \tag{14.61}$$

$$\epsilon_{\alpha\alpha} = \epsilon_{\beta\beta} = 0. \tag{14.62}$$

In a curvilinear system, using velocity components tangential to the characteristics, we get

$$D_\alpha v_\alpha = 0, \quad D_\beta v_\beta = 0, \tag{14.63}$$

where D is the covariant differential operator.

These can also be expressed as

$$dv_\alpha = v_\beta d\theta, \quad dv_\beta = -v_\alpha d\theta, \tag{14.64}$$

along the α line and the β line, respectively. This form of the incompressibility conditions is known as the Geiringer equations. Components of velocity normal to a characteristic curve should be continuous across it; the tangential component may be discontinuous. This possible sliding of the material along a characteristic in plastic flow is the reason why the characteristics are called slip lines.

14.4.3 Example: Symmetric External Cracks

Consider a bar in plane strain with two cracks at the onset of yielding between the two crack tips. A possible slip-line field is shown in Fig. 14.10.

This example illustrates some of the important features of the orthogonal network formed by the slip-line field. Outside the orthogonal network, the material has maximum shear stresses below the yield value. Between the two crack tips, the maximum shear stress is at its yield value $\tau = \sigma_0/2$. Symmetry with respect to the plane of the two cracks establishes the direction of the slip lines at 45°, as shown in the figure. The crack surfaces are traction free with one principal stress, normal to the surface, being zero and the other parallel to the surface. Again, along the line AO the maximum shear planes are at 45° and $\sigma = \pm\sigma_0/2$. We choose the positive sign

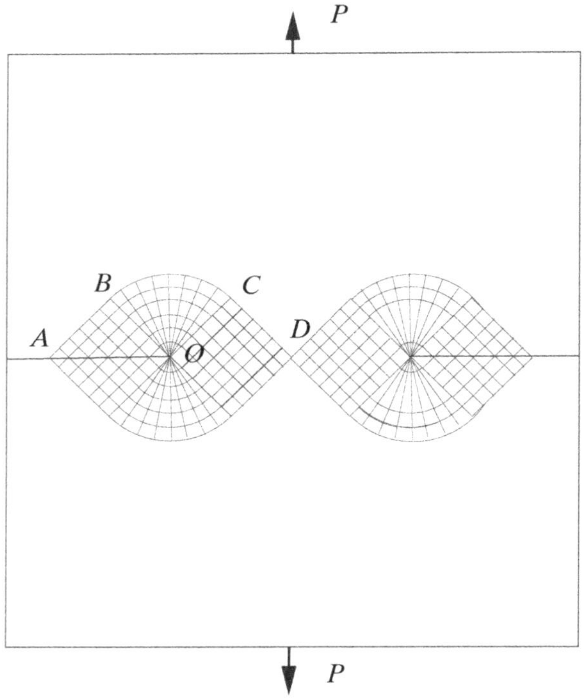

Figure 14.10. Slip-line field for a bar with symmetric cracks.

for the solid under tensile loading. Inside the triangle ABO, the slip lines are straight lines with the α lines at 45° and the β lines at 135°. Inside this triangle the value of σ is constant. There is another constant-stress triangle, OCD, with unknown mean stress, say, σ^*. In between the two triangles there is the region BCO, with O being the crack tip. The crack tip is a singular point as the mean stress jumps from its value in ABO to its value in CDO. However, away from the tip, the mean stress continuously changes along the circular arcs in the area BCO. This area is called a **fan**. Most slip-line fields involving straight boundaries are made of constant-stress triangles and fans. To relate the constant mean stresses in the two triangles, we follow the characteristic line $ABCD$. From B to C the angle θ changes by $-\pi/2$, and, using this in Eq. (14.52), we see

$$\frac{\sigma_B}{\sigma_0} = \frac{\sigma_C}{\sigma_0} - \frac{\pi}{2} \quad \text{or} \quad \sigma_C = \sigma_D = \sigma^* = \sigma_0\left(\frac{\pi+1}{2}\right). \tag{14.65}$$

Along OD, if the horizontal and vertical stresses are σ_{11} and σ_{22}, we have

$$\sigma_{22} - \sigma_{11} = \sigma_0, \qquad \sigma_{22} + \sigma_{11} = \sigma_0(\pi + 1). \tag{14.66}$$

From these equations,

$$\sigma_{22} = \frac{\pi+2}{2}\sigma_0, \qquad \sigma_{11} = \frac{\pi}{2}\sigma_0. \tag{14.67}$$

Note that these results are at variance with the common approximations,

$$\sigma_{22} = \sigma_0, \qquad \sigma_{11} = 0. \tag{14.68}$$

Without going into detail, the velocity distributions are such that the block CDO moves to the right as a rigid body with a velocity of, say, U, and the block ABO has a horizontal velocity of U and a vertical velocity of $2U$. Inside the fan, the velocity components are functions of the angle.

In many cases, slip-line field construction is not unique. There is an underlying assumption that outside the field the stresses satisfy the equilibrium equations with the maximum shear stress below the yield value. There are virtual work principles distinguishing the best approximation among multiple approximations to the velocity fields.

Books on plasticity cited at the end of this chapter have explicit solutions using the slip-line field for many metal-processing cases.

14.5 Drucker's Definition of Stability

In a one-dimensional plot of s against the total deviatoric strain e, a work-hardening constitutive relation would appear as shown in Figure 14.11.

The incremental plastic work done on the body is given by

$$dW^p = s\,de^p. \tag{14.69}$$

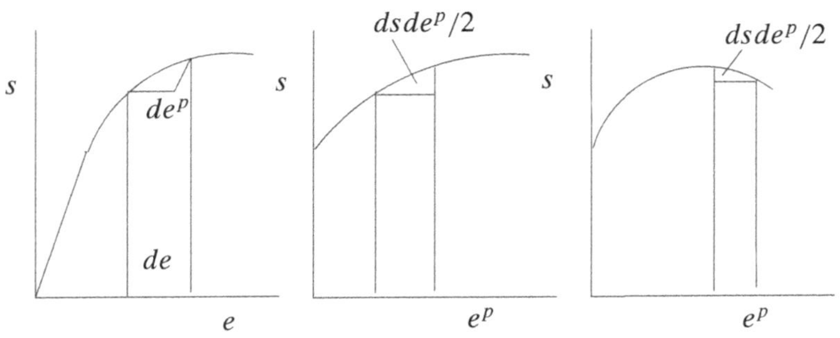

Figure 14.11. Work hardening/softening and plastic work.

We have

$$d^2W^p = dsde^p/2. \tag{14.70}$$

The material is said to be stable if $d^2W^p > 0$. This is known as Drucker's inequality for stable plastic materials. For perfectly plastic materials, we have neutral stability because $d^2W^p = 0$. If we take a point s_1 on the stress–strain curve and another point s_0 in the interior in a partially unloaded state, for stable materials,

$$(s_1 - s_0)de^p > 0. \tag{14.71}$$

Let us consider a cyclic loading situation (see Fig. 14.12), starting from an equilibrium, elastic state $\boldsymbol{s}_0$ at $t = 0$. At $t = \tau_1$, we are at $\boldsymbol{s}_1$ on the yield surface. At $t = \tau_2$, the stress state is $\boldsymbol{s}_2$ outside the original yield surface. In this process the yield surface has expanded because of strain hardening. Subsequent unloading leads to $\boldsymbol{s}_0$ at time τ_3. With

$$de_{ij} = de^e_{ij} + de^p_{ij}, \tag{14.72}$$

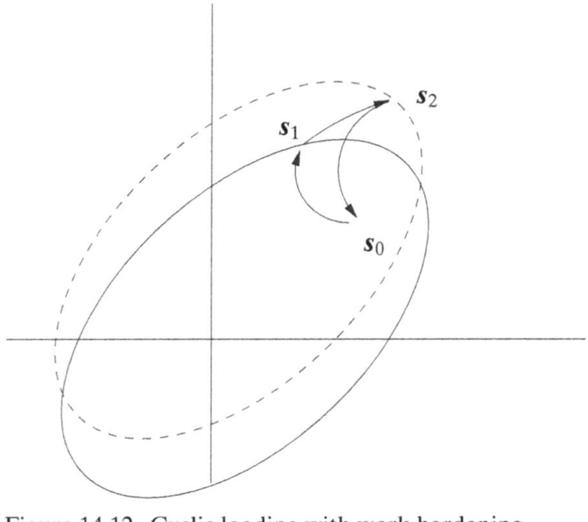

Figure 14.12. Cyclic loading with work hardening.

the work done in this cyclic process is given by

$$\begin{aligned}\Delta W &= \int_0^{\tau_1} s_{ij}\dot{e}^e_{ij}\,dt + \int_{\tau_1}^{\tau_2} s_{ij}(\dot{e}^e_{ij} + \dot{e}^p_{ij})dt + \int_{\tau_2}^{\tau_3} s_{ij}\dot{e}^e_{ij}\,dt \\ &= \oint s_{ij}\dot{e}^e_{ij}\,dt + \int_{\tau_1}^{\tau_2} s_{ij}\dot{e}^p_{ij}dt \\ &= \int_{\tau_1}^{\tau_2} s_{ij}\dot{e}^p_{ij}dt. \end{aligned} \tag{14.73}$$

The work done by the initial loading, if held constant, is written as

$$\Delta W_0 = \int_{\tau_1}^{\tau_2} s_{0ij}\dot{e}^p_{ij}\,dt. \tag{14.74}$$

The work done by an external agency is the difference between ΔW and ΔW_0:

$$\Delta W_{\mathrm{ex}} = \Delta W - \Delta W_0 = \int_{\tau_1}^{\tau_2} (s_{ij} - s_{0ij})\dot{e}^p_{ij}\,dt. \tag{14.75}$$

For this to be nonnegative, the integrand must be nonnegative. That is,

$$(s_{ij} - s_{0ij})\dot{e}^p_{ij} > 0. \tag{14.76}$$

In the limit when s_{0ij} is on the yield surface,

$$\dot{s}_{ij}\dot{e}^p_{ij} > 0. \tag{14.77}$$

On the yield point $\boldsymbol{s}$, if we draw the plastic strain rate $\dot{\boldsymbol{e}}^p$, we must have

$$(\boldsymbol{s} - \boldsymbol{s}_0)\cdot\dot{\boldsymbol{e}}^p > 0. \tag{14.78}$$

The yield surface must be convex for this to be true for all $\boldsymbol{s}_0$. A second conclusion that follows from this is that, if $\dot{\boldsymbol{e}}^p$ is not known, $\dot{\boldsymbol{e}}^p$ must be normal to the yield surface in order for Eq. (14.78) to be true for all $\boldsymbol{s}_0$. Now

$$d\boldsymbol{e}^p = d\lambda\nabla f. \tag{14.79}$$

Thus the Prandtl–Reuss flow rule is the associated flow rule for the von Mises condition. When Eq. (14.79) is expressed as

$$de^p_{ij} = d\lambda\partial f/\partial s_{ij}, \tag{14.80}$$

we have e^p_{ij} derivable from a potential, which is also the yield function. Nonassociative flow rules in which plastic strain components are derivative of a potential separate from the yield function are also found in the literature (Lubliner, 1990).

14.6 Ilíushin's Postulate

In view of the fact that, in perfect plasticity, stress cannot be changed arbitrarily, the use of a strain space yield condition has been recommended by many authors. Extending the single internal variable k to multiple internal variables represented by a vector $\boldsymbol{q}$, the strain space yield condition reads

$$f(\boldsymbol{e}, \boldsymbol{q}) = 0. \tag{14.81}$$

Ilíushin's postulate states that, for any closed curve in strain space,

$$\oint \sigma_{ij} d\epsilon_{ij} \geq 0, \tag{14.82}$$

where the equality holds if the process is entirely elastic.

14.7 Work-Hardening Rules

The manner in which the yield function evolves with plastic deformation can be described by a number of rules.

14.7.1 Perfectly Plastic Material

For this class of materials, it is assumed that the yield function does not change with deformation. From Eq. (14.79)

$$(d\lambda)^2 = d\boldsymbol{e}^p \cdot d\boldsymbol{e}^p / |\nabla f|^2. \tag{14.83}$$

For the von Mises criterion

$$f(\boldsymbol{s}) = \frac{1}{2} s_{ij} s_{ij} - \frac{1}{3}\sigma_0^2 = 0,$$

$$|\nabla f|^2 = s_{ij} s_{ij} = \frac{2}{3}\sigma_0^2,$$

$$d\lambda = \sqrt{3 J_2(de^p)}/\sigma_0. \tag{14.84}$$

Then Eq. (14.80) reads

$$de_{ij}^p = \sqrt{3 J_2(de^p)} s_{ij}/\sigma_0. \tag{14.85}$$

14.7.2 Isotropic Hardening

If the yield surface is assumed to maintain its shape while its size changes with plastic deformation, we then have isotropic hardening. We note that, in the case of cyclic loading, this assumption implies an increased yield strength on the compressive portion of the tension-compression loading. This is opposite to the observed Bauschinger effect. A sketch of the yield surfaces in isotropic hardening is shown in Fig. 14.13. To generalize the uniaxial tensile test results, it is customary to define an effective stress $\bar{\sigma}$ as

$$\bar{\sigma} = \sqrt{3 J_2(\sigma)} = \sqrt{3 s_{ij} s_{ij}/2}, \tag{14.86}$$

which reduces to σ_{11} in the one-dimensional case. The effective plastic strain increment is defined as

$$d\bar{e}^p = \sqrt{4 J_2(de^p)/3} = \sqrt{2 de_{ij}^p e_{ij}^p/3}, \tag{14.87}$$

which reduces to de_{11}^p in the one-dimensional case.

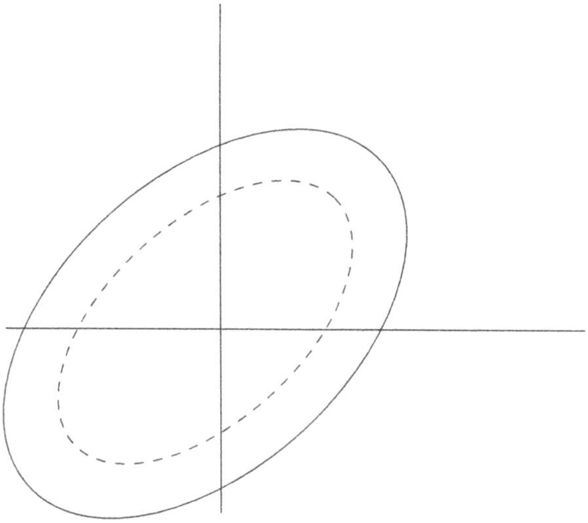

Figure 14.13. Growth of the yield surface in isotropic hardening.

The uniaxial test data can be written as

$$\bar{\sigma} = H\left(\int d\bar{e}^p\right) \quad \text{or} \quad d\bar{\sigma} = h\left(\int d\bar{e}^p\right) d\bar{e}^p. \tag{14.88}$$

The flow rule reads

$$de_{ij}^p = d\lambda s_{ij}, \quad d\lambda = 3d\bar{\sigma}/2\bar{\sigma}h. \tag{14.89}$$

This version of the isotropic-hardening rule is also known as the strain-hardening theory. A second version assumes that the hardening depends on the plastic work done. That is

$$\bar{\sigma} = G(W^p), \tag{14.90}$$

where

$$W^p = \int s_{ij} de_{ij}^p. \tag{14.91}$$

14.7.3 Kinematic Hardening

A hardening rule in which the yield surface does not change in either shape or size but simply translates in the direction of the normal at the current stress state was introduced by Prager in 1955 (see Fig. 14.14). The initial yield surface

$$f(\boldsymbol{s}) = 0 \tag{14.92}$$

is changed, during plastic flow, into

$$f(\boldsymbol{s} - \boldsymbol{\alpha}) = 0. \tag{14.93}$$

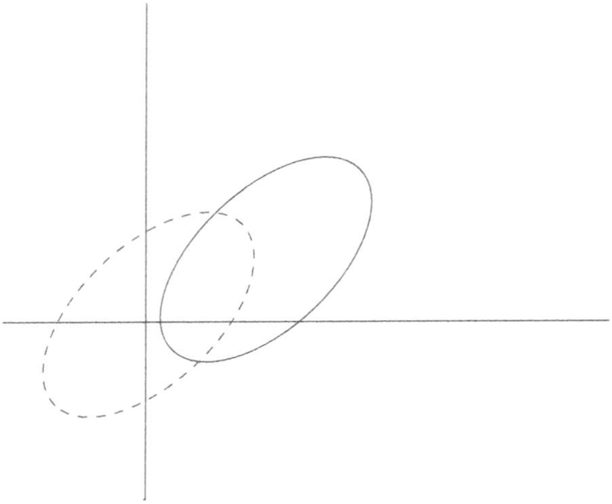

Figure 14.14. Translation of the yield surface in kinematic hardening.

Although originally it was proposed that the incremental change in the rigid body translation is

$$d\boldsymbol{\alpha} = c\, d\mathbf{e}^p, \tag{14.94}$$

this relation leads to some confusion in the six-dimensional stress spaces. In modern applications, Ziegler's modification in the form

$$d\boldsymbol{\alpha} = d\mu(\boldsymbol{s} - \boldsymbol{\alpha}) \tag{14.95}$$

is always used.

To determine the scalar quantity $d\mu$, we note that, from Eq. (14.93),

$$\frac{\partial f}{\partial s_{ij}} ds_{ij} + \frac{\partial f}{\partial \alpha_{ij}} d\alpha_{ij} = 0, \tag{14.96}$$

as the shape of f does not change. From Eq. (14.93) we also have

$$\frac{\partial f}{\partial s_{ij}} = -\frac{\partial f}{\partial \alpha_{ij}}. \tag{14.97}$$

From these,

$$\nabla f \cdot (d\mathbf{s} - d\boldsymbol{\alpha}) = 0, \tag{14.98}$$

$$\nabla f \cdot d\mathbf{s} = d\mu\, \nabla f \cdot (\mathbf{s} - \boldsymbol{\alpha}), \tag{14.99}$$

$$d\mu = \frac{\nabla f \cdot d\boldsymbol{s}}{\nabla f \cdot (\boldsymbol{s} - \boldsymbol{\alpha})}. \tag{14.100}$$

Two other recent theories of hardening are the Mröz theory and the sublayer theory. No attempt is made here to describe these theories.

14.7.4 Hencky's Deformation Theory

The plasticity theories we have discussed so far deal with incremental plastic deformation. These are also called *incremental* or *flow* theories of plasticity. In the deformation or total strain theory, the strain is expressed as a function of the stress as in nonlinear elasticity for loading situations. A linear relation is used for unloading. When the loading is proportional, that is, all components of the stress tensor are proportionally increased, the results obtained with this simpler theory agree with those obtained with the incremental theory. Using the infinitesimal strain measures

$$\epsilon_{ij} = \epsilon^e_{ij} + \epsilon^p_{ij}, \tag{14.101}$$

we have

$$e^e_{ij} = \frac{s_{ij}}{2G}, \qquad \epsilon^e_{ii} = \frac{1}{3K}\sigma_{ii}, \tag{14.102}$$

$$e^p_{ij} = \phi s_{ij}, \quad \text{when} \quad f(\boldsymbol{s}) = 0, \tag{14.103}$$

where

$$\phi^2 = 3\bar{e^p}/2\bar{\sigma}, \quad \bar{e^p} = \sqrt{4J_2(e^p)/3}, \quad \bar{\sigma} = \sqrt{3J_2(s_{ij})}. \tag{14.104}$$

For von Mises criterion, with Prandtl–Reuss equations for the flow rule, if the loading is proportional, it is possible to show that Hencky's theory is identical to the Prandtl–Reuss theory.

14.8 Endochronic Theory of Valanis

In most metals it is difficult to determine the precise location of the yield point on the stress–strain diagram. In monotonic loading a small error in the yield stress may not affect the end result. However, when cyclic loadings are involved, the yield stress along with the hardening rules could lead to large errors in the end result. Also, the time-independent plasticity effects are hard to separate from viscoelastic effects. In 1971, Valanis proposed a unified theory of viscoplasticity without a yield surface. The basis of this theory is an intrinsic "time" measure that governs the inelastic deformation of a neighborhood. This intrinsic time depends on deformation, and it must be an increasing function of strain. We define

$$d\xi = \sqrt{d\epsilon_{ij} d\epsilon_{ij}} \tag{14.105}$$

as an intrinsic measure of strain. When real-time dependence is involved, we define an endochronic time scale

$$d\zeta = \sqrt{\alpha^2 d\xi^2 + \beta^2 dt^2}, \tag{14.106}$$

where α and β are material constants. A monotonically increasing material function z is then introduced as

$$z = z(\zeta), \qquad z' > 0. \tag{14.107}$$

Whenever the body is subjected to strain, z will be increasing. The stress–strain relations are given by

$$s_{ij} = \int_{z_0}^{z} 2\mu(z-z')\frac{\partial e_{ij}}{\partial z'}dz', \tag{14.108}$$

$$-p = \int_{z_0}^{z} [K(z-z')\frac{\partial \epsilon_{kk}}{\partial z'} + D(z-z')\frac{\partial \theta}{\partial z'}]dz', \tag{14.109}$$

where θ is the coefficient of thermal expansion and

$$2\mu = 2\mu_0 H(z) + 2\mu_1 e^{-\alpha_1 z} + 2\mu_2 e^{-\alpha_2 z} + \cdots +, \tag{14.110}$$

$$K = K_0 H(z) + 2K_1 e^{-\alpha_1 z} + 2K_2 e^{-\alpha_2 z} + \cdots +, \tag{14.111}$$

$$D = D_0 H(z) + 2D_1 e^{-\alpha_1 z} + 2D_2 e^{-\alpha_2 z} + \cdots +, \tag{14.112}$$

and $\mu_0, \mu_1, \ldots, K_0, K_1, \ldots,$ and $D_0, D_1, \ldots$ are constants.

If the material is plastically incompressible we have

$$-p = K\epsilon_{kk} \tag{14.113}$$

in place of Eq. (14.109). If μ contains only a single term of the form

$$2\mu = 2\mu_1 e^{-\alpha_1 z}, \tag{14.114}$$

we have

$$s_{ij} = 2\mu_1 \int_{z_0}^{z} e^{-\alpha_1(z-z')} de_{ij}(z'), \tag{14.115}$$

$$\frac{ds_{ij}}{dz} = 2\mu_1 \frac{de_{ij}}{dz} - \alpha_1 s_{ij}, \tag{14.116}$$

$$de_{ij} = \frac{1}{2\mu_1} ds_{ij} + \frac{\alpha_1 dz}{2\mu_1} s_{ij}. \tag{14.117}$$

If we divide the strains into elastic and inelastic parts we have

$$de_{ij} = de_{ij}^e + de_{ij}^p. \tag{14.118}$$

Then

$$de_{ij}^p = \frac{\alpha_1 dz}{2\mu_1} s_{ij}, \qquad de_{ij}^e = \frac{ds_{ij}}{2\mu_1}. \tag{14.119}$$

The first of these is the Prandtl–Reuss equation of plasticity. Also

$$\frac{\alpha_1 dz}{2\mu_1} = \sqrt{\frac{J_2(de^p)}{J_2(s)}}, \tag{14.120}$$

$$dz = \frac{2\mu_1}{\alpha_1}\sqrt{\frac{J_2(de^p)}{J_2(s)}}, \tag{14.121}$$

$$de_{ij}^p = \frac{\sqrt{J_2(de^p)}}{J_2(s)}. \tag{14.122}$$

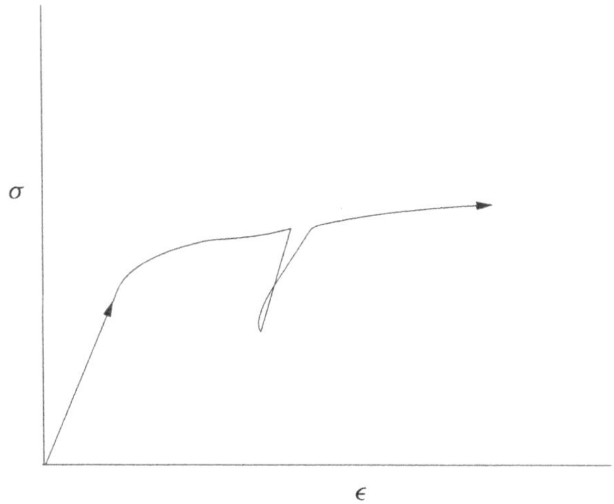

Figure 14.15. Closed loop in a viscoplastic stress–strain diagram.

The original endochronic theory just described has been subjected to a series of criticisms. One of these deals with the predictions of the theory concerning the behavior under unloading–reloading. The theory violates Drucker's postulate of material stability. Because Drucker's postulate is not based on thermodynamics, this criticism is not considered relevant. A more serious criticism has to do with the open loop predicted by the theory in small amounts of unloading and reloading. Experiments indicate a closed loop, as shown in Fig. 14.15.

To overcome this inadequacy, Valanis proposed a modified theory in 1980. For simplicity, we confine ourselves to a version of the modified theory applicable to plastically incompressible materials.

The intrinsic time measure is defined with the plastic strain increments as

$$d\xi^2 = de^p_{ij} e^p_{ij}, \tag{14.123}$$

instead of the total strain increments. The deviatoric stresses are given by the integral representation

$$s_{ij} = \int_0^z \rho(z - z') \frac{\partial e^p_{ij}}{\partial z'} dz', \tag{14.124}$$

where, to have the material yield at the onset of loading, it is necessary to have

$$\rho(0) \to \infty. \tag{14.125}$$

The modified theory proposed by Valanis also incorporates plastic compressibility and coupling between dilatation and distortion.

14.9 Plasticity and Damage

The characterization of damage in materials has been an ongoing effort in the general area of continuum mechanics. In polycrystalline materials, one observes a high concentration of dislocations and grain boundary sliding at the microscopic levels.

In brittle solids these singularities are replaced with microcracks. Applied stresses at the macroscopic levels are the driving forces responsible for the movement of dislocations and the extension of microcracks—phenomena that are, at least partially, irreversible. The cumulative effect of these irreversibilities is the deterioration of the strength of the material and eventual failure under repeated loadings.

Rice (1971, 1975) introduced an internal variable approach to treat solids undergoing irreversible thermodynamic processes. The internal variables are supposed to describe the microcrack extensions, dislocation movements, etc. It is important to note that internal variables are not directly observable, unlike macroscopic variables: strain, stress, and temperature. Our subsequent discussion follows the work of Yang, Tham and Swoboda (2005), which specializes the results obtained by Rice by assuming homogeneous functional forms for certain quantities. Although the context of the original papers by Rice was irreversible deformations in plasticity, we adopt the formalism for a general treatment of damage in materials. A key assumption of the theory is that the increments of internal kinematic variables such as microcrack lengths are driven by associated thermodynamic microforces that are functions of macroforces, namely, the applied stresses, and these increments do not depend on the external stresses directly.

Consider a unit volume of a material sample in which the macrovariables, strain, stress, and temperature, are uniform and the collection of internal variables q_i $(i = 1, 2, \ldots, n)$ are denoted by H. These variables are *not state variables* in the thermodynamic sense, and the damage at any time is history dependent. Helmholtz free energy U and Gibbs free energy G are expressed as

$$U = U(\boldsymbol{e}, T, H), \quad G = G(\boldsymbol{\sigma}, T, H), \tag{14.126}$$

where $\boldsymbol{e}$ and $\boldsymbol{\sigma}$ are strain and its conjugate stress, respectively, and T is the temperature. At a fixed value of H, we have the elastic relations

$$\boldsymbol{\sigma} = \frac{\partial U}{\partial \boldsymbol{e}}, \quad \boldsymbol{e} = -\frac{\partial G}{\partial \boldsymbol{\sigma}}. \tag{14.127}$$

As mentioned earlier, a change in H corresponds to changes in q_i, and associated with each increment dq_i, there is a microforce X_i. The balance of energy states

$$\begin{aligned} X_i dq_i &= -\frac{\partial U}{\partial H} dH \equiv -dU^P \\ &= -\frac{\partial G}{\partial H} dH \equiv -dG^P. \end{aligned} \tag{14.128}$$

The driving forces X_i are functions of the macrovariables and the extent of damage

$$X_i = X_i(\boldsymbol{e}, T, H), \tag{14.129}$$

where dependence on H is through the history of all the q_i's. The irreversible entropy production rate

$$\dot{S}_I = \frac{1}{T} X_i \dot{q}_i \geq 0. \tag{14.130}$$

We introduce a potential $B(X_i, T, H)$ to define the direction of the damage flow:

$$\dot{q}_i = \frac{\partial B}{\partial X_i}, \quad B = \int^X \dot{q}_i dX_i, \tag{14.131}$$

where we have used vector notation for the n-dimensional quantities X_i, and the integration has to be carried out with fixed T and H. These are the evolution equations for the kinetics of the damage variables. The function B may be highly nonlinear. The direction of the vector $\dot{\boldsymbol{q}}$ is along the normal to the surface B.

The inelastic strain increment can be written as

$$\begin{aligned} d\boldsymbol{e}^P &= \boldsymbol{e}(\boldsymbol{\sigma}, T, H + dH) - \boldsymbol{e}(\boldsymbol{\sigma}, T, H) \\ &= -\frac{\partial^2 G}{\partial \boldsymbol{\sigma} \partial H} dH \\ &= -\frac{\partial dG^P}{\partial \boldsymbol{\sigma}} \\ &= \frac{\partial X_i}{\partial \boldsymbol{\sigma}} dq_i. \end{aligned} \tag{14.132}$$

The strain rate becomes

$$\dot{\boldsymbol{e}}^P = \frac{\partial X_i}{\partial \boldsymbol{\sigma}} \dot{q}_i = \frac{\partial B}{\partial X_i} \frac{\partial X_i}{\partial \boldsymbol{\sigma}} = \frac{\partial B}{\partial \boldsymbol{\sigma}}. \tag{14.133}$$

The dissipation rate $D(X_i, T, H)$ is defined as

$$D = X_i \dot{q}_i = X_i \frac{\partial B}{\partial X_i} = -\frac{\partial U}{\partial q_i} \dot{q}_i = -\left.\frac{\partial U}{\partial t}\right|_{T,e}. \tag{14.134}$$

At this point we make two assumptions about the potential B:

1. B is a homogeneous function of degree r in its variables X_i.
2. B is also a homogeneous function of degree s in its variables σ_{ij}.

Our first assumption is more general than the one used in Yang et al. (2005); they assume each $\dot{q}_i$ is a homogeneous function of all X_j. As a consequence of Euler's theorem concerning homogeneous functions,

$$X_i \dot{q}_i = X_i \frac{\partial B}{\partial X_i} = rB = D. \tag{14.135}$$

As D is positive semidefinite, so is B. From differentiating

$$D = X_i \frac{\partial B}{\partial X_i}, \tag{14.136}$$

we find

$$\begin{aligned} r\dot{q}_i &= \dot{q}_i + X_j \frac{\partial^2 B}{\partial X_i \partial X_j}, \\ \dot{q}_i &= J_{ij} X_j, \end{aligned} \tag{14.137}$$

where J_{ij} are the Onsager coefficients relating fluxes and forces, obtained here as

$$J_{ij} = \frac{1}{r-1}\frac{\partial^2 B}{\partial X_i \partial X_j}. \tag{14.138}$$

These coefficients come out to be symmetric because of the symmetry of the mixed partial derivatives. Also, we need $r \geq 2$.

As a consequence of the second assumption, Eq. (14.133) gives

$$\sigma_{ij}\dot{e}^P_{ij} = sB = \frac{s}{r}D, \tag{14.139}$$

which is positive semidefinite.

We note that it has been recognized that all the internal variables are difficult to account for and, in all practical cases, they have to be replaced with a certain average or equivalent set of a smaller number of variables.

14.10 Minimum Dissipation Rate Principle

Instead of prescribing a flow rule as in Eq. (14.131), we may use the the principle of minimum dissipation rate to derive the flow rule.

For a given power input for creating damage,

$$X_i\dot{q}_i = k, \tag{14.140}$$

we may ask: What values of X_i do we need to have the dissipation a local minimum?

For this extremum problem, we set up a modified function using a Lagrange multiplier λ as

$$D^* = D - \lambda(X_i\dot{q}_i - k). \tag{14.141}$$

Minimization of D^* leads to

$$\lambda\dot{q}_i = \frac{\partial D}{\partial X_i}. \tag{14.142}$$

These relations show that $\dot{\boldsymbol{q}}$ is directed normal to the surface D. If we further impose the condition that D is a homogeneous function of degree r, multiplying Eq. (14.142) by X_i, we get

$$\lambda X_i\dot{q}_i = X_i\frac{\partial D}{\partial X_i} = rD_{\min}. \tag{14.143}$$

Then, $\lambda = r$. As D is a highly nonlinear function, unlike a quadratic, we may have multiple local minima. Often a condition of convexity is imposed on D to avoid this. This minimum principle uses the generalized forces as variables. If we use the complement of D with the conjugate variables $\dot{q}_i$, we get a maximum dissipation principle equivalent to the choice of strain rate direction with respect to a yield surface in plasticity.

The application of these ideas to damage is still at its infancy. The paper by Krajcinovic (2004) is recommended for further reading.

SUGGESTED READING

Hill, R. (1950). *The Mathematical Theory of Plasticity*, Oxford University Press.
Johnson, W. and Mellor, P. B. (1973). *Engineering Plasticity*, Van Nostrand Reinhold.
Kachanov, L. M. (2004). *Fundamentals of the Theory of Plasticity*, Dover.
Krajcinovic, D. (2004). Damage mechanics: Accomplishments, trends and needs, *Int. J. Solids Struct.*, **37**, 267–277.
Lubliner, J. (1990). *Plasticity Theory*, Macmillan.
Nemat-Nasser, S. (2004). *Plasticity: A Treatise on Finite Deformation of Heterogeneous Inelastic Materials,* Cambridge University Press.
Prager, W. (1955). The Theory of Plasticity: A Survey of Recent Achievements, *Proc. Instn. Mech. Engrs.*, 41–57.
Rajagopal, K. R. and Srinivasa, A. R. (2004). On thermo-mechanical restrictions of continua, *Proc. R. Soc. London Ser. A*, **460**, 631–651.
Rice, J. R. (1971). Inelastic constitutive relations for solids: An integral variable theory and its applications to metal plasticity, *J. Mech. Phys. Solids*, **19**, 433–455.
Rice, J. R. (1975). Continuum mechanics and thermodynamics of plasticity in relation to microscale deformation mechanisms, in *Constitutive Equations in Plasticity* (A. S. Argon, ed.), MIT Press, 23–79.
Spitzig, W. A. and Richmond, O. (1984). The effect of pressure on the flow stress of metals, *Acta Metal.*, **32**, 457–463.
Ulm, F.-J. and Coussy, O. (2003). *Mechanics and Durability of Solids*, Prentice-Hall, Vol. 1.
Valanis, K. C. (1971). A theory of viscoplasticity without a yield surface, Part I. Theory, *Arch. Mech.*, **23**, 517–533.
Valanis, K. C. (1971). A theory of viscoplasticity without a yield surface, Part II. Application to mechanical behavior of metals, *Arch. Mech.*, **23**, 535–551.
Valanis, K. C. (1980). Fundamental consequence of a new intrinsic time measure plasticity as a limit of the endochronic theory, *Arch. Mech.*, **32**, 171–191.
Yang, Q., Tham, L. G., and Swoboda, G. (2005). Normality structures with homogeneous kinetic rate laws, *J. Appl. Mech.*, **72**, 322–329.
Ziegler, H. (1959). A Modification of Prager's Hardening Rule, *Quat. Appl. Math.*, **17**, 55–65.

EXERCISES

14.1. A thin-walled tube with circular cross-section is subjected to combined tension and torsion. If ℓ, r, t, and θ denote, respectively, the current length, mean radius, thickness, and angle of twist of the tube, determine the infinitesimal-strain increments in a further deformation defined by $\delta\ell$, δr, δt, and $\delta\theta$.

14.2. A rod is extended in simple tension by gradually increasing the tensile stress σ. If the yield stress is a linear function of the physical longitudinal strain ϵ, with slope h, determine the plastic work done in producing the strain ϵ beyond the yield limit in terms of h, Young's modulus E, and the initial yield stress σ_0.

14.3. A metal plate occupies the region $-a < x < a$, $-b < y < b$, and $-t < z < t$. It is subjected to proportional loading along its two axes by increasing the parameter α, with $\sigma_{xx} = 2\alpha$, and $\sigma_{yy} = \alpha$. All other stress components are zero. If yielding is governed by von Mises condition, obtain the stresses at yield. If an additional plastic strain of $\epsilon_{xx}^p = 0.1$ is imparted through the same proportional loading, obtain the plastic strains in the other directions, assuming incompressibility of plastic strains.

14.4. A thin-walled circular cylindrical tube of elastic-plastic strain-hardening material is stressed in simple tension to the point of yielding and then, holding the axial stress constant, it is twisted by a continuously increasing end torque. If the strain-hardening law is

$$\sqrt{3J_2} = \sigma_0 \left[1 + \frac{3}{4}\bar{\epsilon^p} \right],$$

where σ_0 is the initial yield stress and the material satisfies the von Mises yield condition and the associated flow rule, obtain the relative angle of twist.
Let L, R, and G represent the initial length, radius of the tube, and the shear modulus, respectively.

14.5. Consider a thin-walled tube of mean radius R, length L, and thickness t, made of a rigid plastic hardening material. We obtain the hardening rule from uniaxial experiments by fitting the curve

$$\bar{\sigma} = \sigma_0[0.99 + \sqrt{0.0001 + \bar{\epsilon}}],$$

where σ_0 is the initial yield stress and $\bar{\sigma}$ and $\bar{\epsilon}$ are effective stress and strain, respectively. Assume von Mises yield condition and the associated flow rule.

The tube is first subjected to a shear stress of $\sigma_0/\sqrt{3}$ through an applied torque. Keeping the shear stress constant, a tensile load $\sigma \ll \sigma_0$ is applied. Obtain the percentage changes in the dimensions of the tube.

14.6. Assume the thin-walled tube in the previous exercise is made of an elastic, isotropic linear hardening material with the uniaxial stress–strain relation

$$\sigma = E\epsilon, \quad \sigma < \sigma_0, \quad \sigma = (1 - \frac{\alpha}{E})\sigma_0 + \alpha\epsilon, \quad \sigma \geq \sigma_0.$$

Using the von Mises condition and associative flow rule for yielding and Poisson's ratio ν, convert the uniaxial law to a relation between effective stress and strain.

The tube is twisted to a point of initial yield and then, keeping the torque constant, it is stretched to a length of $(1 + \beta)L$. What are the axial stress and the angle of twist at this point? Next, the torque is removed and then the axial tension is reduced to zero. What are the final dimensions of the tube?

14.7. A tensile test specimen with initial length L_0 and cross-sectional area A_0 is made from an incompressible rigid hardening plastic material. The plastic strain is related to the true stress in the form

$$\epsilon = \int_{L_0}^{L} \frac{dL}{L} = K\sigma^n,$$

where K and n are constants. Compute the force P required to have a displacement U. Sketch P as a function of U and find the location of the maximum load. What is the physical significance of a maximum load?

14.8. The Stéfani model for plasticity (see Fig. 14.16) involves a spring with stiffness E in series with two friction and spring elements (from Ulm and Coussy, 2003). Obtain the load displacement diagram for this model during a complete displacement cycle: $0 \to U \to 0 \to -U \to 0$. Obtain an expression for the energy dissipated in

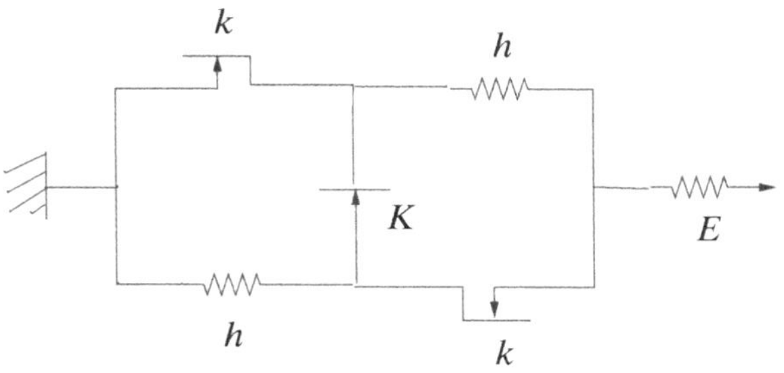

Figure 14.16. The Stéfani model for one-dimensional plasticity.

the process. What is the constraint on the constants in order to have positive dissipation? Assume $K > 2k$.

14.9. A cantilever beam of rectangular cross-section has height $2c$ and length L. Under an applied load P at the end $x = L$, the longitudinal strain is given by

$$\epsilon = \kappa y,$$

where κ is the curvature of the neutral axis and $-c < y < c$. Assuming Young's modulus is E and the material is elastic perfectly plastic with a yield stress of σ_0, obtain the load P_p at the onset of plastic flow. What is the load P_c when the cantilever collapses? Obtain the equation $y = f(x)$, describing the elastic-plastic boundary when $P_p < P < P_c$. If the load is removed after the initial yield, sketch the distribution of the residual stress.

14.10. For a rigid plastic material in plane strain, if the yield stress τ depends on the mean stress $\sigma = (\sigma_1 + \sigma_2)/2$, show that the condition for the equilibrium equations to be hyperbolic at yield is $d\tau/d\sigma < 1$. Derive the equations for the characteristic curves and see if they are orthogonal.

14.11. Consider a rigid plastic tensile specimen in plane strain with two symmetric, sharp notches with angle γ, instead of the sharp crack problem described in this chapter. Obtain the slip-line field and the stress distributions in the yielded regions, using the constant-stress triangles and fans.

14.12. A wedge with an apex angle of 2γ is in plane strain. On one side of the wedge a pressure p is applied to a distance of a from the tip. Construct slip-line fields by using constant-stress triangles and fans to obtain the value of p when the tip begins to yield. Consider $\gamma < \pi/4$ and $\gamma > \pi/4$ separately.

Author Index

Subject Index